DR. SCHUESSLER'S BIOCHEMISTRY

A New Domestic Treatise
A Medical Book for the Home

By J. B. CHAPMAN, M. D.
Edited by J. W. COGSWELL, M. D.

□

The Theory, Action
and
Practical Application
of
Schuessler's
Twelve Biochemic Remedies

B. Jain Publishers Pvt. Ltd.
NEW DELHI-110055

NOTE FROM THE PUBLISHERS

Any information given in this book is not intended to be taken as a replacement for medical advice. Any person with a condition requiring medical attention should consult a qualified practitioner or therapist.

Price: Rs. 60.00

Reprint Edition: 2004

Published by
KULDEEP JAIN
for

B. Jain Publishers (P) Ltd.

1921, Chuna Mandi, St. 10th, Paharganj,
New Delhi-110 055
Phones: 2358 0800, 2358 1100, 2358 1300, 2358 3100, 5169 8991
Fax: 011-2358 0471, 5169 8993; *Email:* bjain@vsnl.com
Website: **www.bjainbooks.com**

Printed in India by
Unisons Techno Financial Consultants (P) Ltd.
522, FIE, Patpar Ganj, Delhi-110 092

ISBN 81-7021-164-6
BOOK CODE B-2567

Foreword

The following pages are intended to be a guide to the home treatment of common diseases by means of the Schuessler Biochemic Remedies.

It has been thought best to include a description of some of the more serious diseases to assist in the recognition of such conditions, in order that the advice of a physician may be secured at once. No home treatment is therefore included under these headings.

J. W. C.

PREFACE

□

RECOGNIZING the fact that Biochemistry is a system of medicine which is adapted to domestic purposes—owing to the use of non-poisonous remedies, the small number of remedies needed, and the rational theory upon which it is based—the author boldly launches this work on its mission to the people, firmly believing that those for whom it is intended will receive it in the same helpful spirit in which it is given.

In the past, a study of medical history will show, man has been groping along the wrong way. Strong drugs, poisons, nauseous chemicals—these were practically the only means employed to overcome disease. Sickness is now regarded as an unnatural condition—one at variance with the intentions of nature—in fact, nature itself is constantly battling against such conditions, and by studying nature we can discover the logical ways of overcoming disease.

Gradually it dawned upon the scientific world that within the body itself are to be found the most potent weapons in the battle against disease and that the natural constituents of the human body are effective remedies in this battle. It is upon this great truth that Dr. Med. W. H. Schuessler has built his system of Biochemistry.

Biochemistry has, since its inception, been received with gladness by the sick and suffering of both the old and new world. From the sultry jungles of India to the fir-clad shores of Puget Sound, come reports of its curative powers.

In the present work I have endeavored to give a simple outline of Biochemistry; one adapted to the needs of the millions. My object has been to divest the subject of technical words and phrases which serve to distract and mystify the average reader. I have aimed to make the arrangement so complete that none may fail to recognize their diseases and adopt the proper course of treatment. Not forgetting the noble band of men and women who devote their lives and talents to the cause of suffering humanity; the physicians who are ever ready to answer duty's call—I have no desire to antagonize them in the publication of this work—on the contrary, it is by educating the people to a knowledge of the time when a physician is necessary, that his services are appreciated.

For those who are deprived of the services of a medical adviser, or who are unable to obtain one promptly in case of sickness, this volume is particularly intended.

Yours in sincerity,

J. B. CHAPMAN, M.D.

The Simple Logic of Dr. Schuessler's Biochemic Theory

□

1. The human body contains twelve vital inorganic elements which are responsible for maintenance of normal cell-function.

2. When from some cause, one or more of these elements become deficient the normal cell-function or metabolism is disturbed and a condition arises known as disease.

3. By supplying to the system the lacking elements in the form of Schuessler Biochemic Remedies normal cell-function and *health* can be restored.

The Schuessler Biochemic Remedies are perfectly safe to take for adults and children.

A Brief Sketch of the Discovery and Development of

The Biochemic System of Treating Disease

originated by

DR. MED. W. H. SCHUESSLER

□

HISTORY

Biochemistry had its birth in 1832, when the following brief statement was written in Stapf's *Archiv: "All the essential component parts of the human body are great remedies."* Another record was made in the same journal in 1846: *"All constituents of the human body act on such organs principally where they have a function."*

At a later period, Grauvogl, in his Text-book, took some notice of those remarks.

In 1873, Dr. Med. Schuessler, of Oldenburg, Germany, published an article entitled: "A Shortened Therapeutics," in which he said: *"About a year ago I intended to find out, by experiments on the sick, if it were not possible to heal them, provided their diseases were curable at all, with some substances that are the natural, i. e., physiological function remedies."*

Dr. Lorbacher, of Leipzig, made some criticisms on this article, and called forth a detailed reply from

Schuessler, which ran through seven numbers of the journal. He denominated the new theory the "*Abridged System of Therapeutics.*"

Dr. H. C. G. Luyties translated Schuessler's original communication into the English language.

Later, Dr. C. Hering, one of the great apostles of the homeopathic school, wrote a small work on the "*Twelve Biochemic Remedies,*" and "*recommended for investigation*" this interesting and valuable discovery. Several editions of the work were published in rapid succession.

The twelfth edition of Schuessler's work was translated by J. T. O'Connor, M.D., and another by M. Docetti Walker, M.D., of Dundee, Scotland, which was considerably enlarged by the addition of an appendix, popularizing the biochemic method.

In the early spring of 1894, "*The Biochemic System of Medicine,*" by Geo. W. Carey, M.D., was published by F. August Luyties, St. Louis, Mo., and has already run through many large editions. All of the above works were written especially for the medical profession and the author believes he is the first to give to *the people a thoroughly domestic treatise* on the subject of Biochemistry.

Since 1832, when the great key-note of Biochemistry was struck ("*all the essential component parts of the human body are great remedies*"), the subject has been thoroughly investigated and endorsed by thousands of broadminded, progressive physicians of the old and new continents.

WHAT IS BIOCHEMISTRY?

The word *biochemistry* is derived from a Greek word—*bios* (meaning life)—and chemistry; therefore, its true meaning is *the chemistry of life.* But usage has given the word a different signification, and the following has been given as a more accurate definition: "That branch of science which treats of the composition of animal and vegetable matter; the process by which the various fluids and tissues are formed; the nature, cause and correction of the abnormal condition called disease."

Professor Moleschott, of Rome, says (and, indeed, it was these words which gave Schuessler his inspiration): "The structure and vitality of the organs *depend upon the presence of the necessary quantities of inorganic constituents.*

"On this fact is based the high estimation in which of late years the subject of the relative preparations of the inorganic substances to the individual parts of the body have been held.

"This estimation neither proudly despises any fact, nor fosters, on the other hand, futile hopes; but promises both to Agriculture and Medicine a brilliant future.

"In the face of such positive facts, it can no longer be denied that the substances which remain after incineration or combustion of the tissues—the ashes—are as important and essential to the inner composition, and consequently to the 'form giving' and 'kind determining' bases of the tissues, as those substances which are volatilized during combustion.

"A glue-furnishing base and bone-earth are essential constituents of bone; without either there can be no true bone; so also there can be no cartilage without cartilage-salt; nor blood without iron; nor saliva without potassium chloride."

The human body is composed of two kinds of matter —*organic* and *inorganic*. The former greatly preponderates, but it does not follow that it is more essential to life than the latter; indeed, the organic could not perform its proper function without the inorganic. These are not mere theories but scientifically proven facts. It has been discovered that the human body will survive for a shorter period of time from the deprivation of inorganic (mineral) salts than of the other (organic) constituents of the diet.

It is upon the relative quantities of these two materials that life and health depend.

An analysis of the blood shows it to contain organic and inorganic matter which is constantly being built into the human structure.

The organic constituents are sugar, fats and albuminous substances. The inorganic constituents are water and certain cell-salts. The relative quantities in the human organism are about as follows: water, seven-tenths; cell-salts, one twentieth, and organic matter the remainder. Being so small in quantity, the cell-salts have, until lately, been thought to be of little importance. But now it is known that they are the vital portion of the body, *the workers, the builders;* that water and organic substances are simply inert matter used by these salts in building the cells of the body.

The twelve inorganic (mineral) salts are all essential to the proper growth and development of every part of the system.

They are:

CALCAREA FLUOR.
(Fluoride of Lime)

CALCAREA PHOS.
(Phosphate of Lime)

CALCAREA SULPH.
(Sulphate of Lime)

FERRUM PHOS.
(Phosphate of Iron)

KALI MUR.
(Chloride of Potash)

KALI PHOS.
(Phosphate of Potash)

KALI SULPH.
(Sulphate of Potash)

MAGNESIA PHOS.
(Phosphate of Magnesia)

NATRUM MUR.
(Chloride of Soda)

NATRUM PHOS.
(Phosphate of Soda)

NATRUM SULPH.
(Sulphate of Soda)

SILICEA
(Silica)

Should a deficiency occur in one or more of these workers, some abnormal or "diseased" conditions arise, and according as they manifest themselves in different ways and in different parts of the body, they have been given various names. *Every disease which afflicts humanity reveals a lack of one or more of these inorganic cell-salts. Health and strength can be maintained only so long as the system is properly supplied with these cell-workers or tissue-builders.*

Dr. Schuessler says: "The *inorganic substances* in the blood and tissues are *sufficient to heal all diseases which are curable at all.* If the remedies are used according to the symptoms, the desired end will be gained by means of the application of natural laws."

Dr. Schuessler's Biochemistry seeks to ascertain what salts are lacking and supply them in just the form needed. Any disturbance in the motion of these cell-salts in living tissues, constituting disease, can be rectified and the equilibrium re-established by administering the same salts in small quantities. This is brought about by virtue of chemical affinity in the domain of histology. Therefore, this therapeutic procedure is styled by Dr. Schuessler, the Biochemic Method, and stress is laid on the fact that *it is in harmony with well-known facts and laws in physiological chemistry and allied sciences.*

It is the blood that contains the material for every tissue of the body, that supplies nutriment to every organ, enabling it to perform its individual function; it is, indeed, *a microcosm,* able to supply every possible want to the animal economy.

When a plant droops for the want of water or for some fertlizing material which is necessary for its growth—and which is a constituent of the plant—*we know* that if the lacking materials are supplied, it will revive and bloom again. It is the same with the human system—it is composed of certain materials, and if they become deficient in quantity or disturbed in their molecular motion, disease and death is the result; to restore the system to its normal, healthy condition, it is only necessary to supply the deficient material, in a form which can be used by the blood.

The author is averse to closing this chapter without adding the following very excellent letter, which appeared in a Medical Journal, feeling that the article would be incomplete were it omitted:

"Without a doubt, *the greatest discovery made in the science of medicine in the past half century* is the Biochemic Theory; based, as it is, on scientific, logical deductions, it seems strange that these truths, so recently revealed, should have remained hidden from the conception of the medical investigator of this advanced era of medical research. While it is true that the theory of supplying deficiencies is, and has been, since its discovery, fought by the ablest, but I cannot say most liberal-minded, advocates of medicine, its principles are so simple that they can readily be grasped by any unbiased, unprejudiced mind.

"What is more rational, what more natural, founded as it is on natural law, that where there be a deficiency in one or more of the component parts of the constituents of an organism, that this deficiency will produce a deranged or a diseased condition; or more logical, than by the replacing of these lacking elements an equilibrium will again be restored and the organism returned to its normal condition? *Chemistry has demonstrated* that the human body is composed of water principally, of organic matter, and of lime, potassium, sodium, iron and magnesium, and that these cell-salts enter into the composition, in their proper proportions, of every tissue of the body.

"Replying to the question: 'What is health?' a noted professor unhesitatingly answered: 'An harmonious relation of all the organs of the body.' I inferred that he meant the functions of the organs of the body must constantly be in harmonious relation. Had he gone a little deeper, he would have explained that the health of each individual organ, *per se,* was dependent upon the harmonious relation of each individual cell of which the organ was composed, and that the activity of each cell was in turn dependent upon its component parts, namely: the elements which in combination with the organic matter formed the cell—*cell-salts.*

"Schuessler solved the problem correctly when he stated that health was dependent upon the proper quantity and distribution of the inorganic materials of the system, and that a deficiency or an unequalization of any of these constituted disease.

"Tissue is not composed alone of the mineral or cell-salts, but of organic substances as well. One-twentieth of the human body is composed of the inorganic salts, the remainder water and organic matter. But the water and organic matter is inert and useless in *the absence* of the inorganic cell-salts. These salts are the *builders,* the workmen, who use the organic matter, albumen, sugar, oil, fibrine, and also water, to build up tissue. Therefore, a deficiency in these workmen will retard the processes of life. Without a proper supply of these builders in the blood, new tissue cannot be supplied as fast as the old decays, and it must be plainly seen that a lack of these workmen is the cause of disease.

"Are the laws of other schools of medicine based upon principles more in unison and accord with Nature's divine law—the supplying of deficiencies? Or, are they the deductions of more logical reasonings than the restoration to an equalization, to an equilibrium, or to harmony, by supplying that which is the direct cause of this disorganization?

"The laws of nature are definite as to exact requirements of the various elements by the different parts of the human body. An abnormal condition or a deficiency of certain elements is manifested by symptoms. These symptoms too many persons believe are the diseases themselves. Correctly speaking, a symptom is merely a sign—a signal of distress, as it were—that Nature gives as a warning that certain parts are endangered, and indicates the elements required for restoration of a normal condition.

"In Biochemistry, then, a symptom is merely indicative of a deficiency of one or more of the cell-salts which compose the tissues involved. Supply this factor, be it lime, iron, magnesia, sodium or potassium, and the reaction will follow promptly, and equality and harmony—health—be established."

THE PROPER PREPARATION OF THE BIOCHEMIC REMEDIES

In order to be assmilated or taken into the substance of the cells of the human body, matter must be reduced to almost infinitesimal size. Nature does this with the inorganic cell-salts that are taken up by plants from the earth in microscopic quantities, introduced into the digestive system with the food which we eat, and carried by the blood stream through the system into the cells, where they are absorbed. This is Nature's method of supplying the cell-salts to the body, but, unfortunately, the supply is not always sufficient.

When disease results from an insufficiency of the inorganic cell-salts, they should be supplied in a manner as closely simulating Nature's method as possible. Dr. Schuessler accomplished this by a special method of triturating the inorganic cell-salts, in this way reducing their particles to a size that would insure their ready assimilation by the cells, and their proper union with the organic matter which they control.

DOSE

The 12 Schuessler Remedies should be used in the "celloid" form. "Celloids" are small, agreeable tablets, convenient to take and always insure correct dosage. "Celloids" of the Schuessler Remedies are perfectly safe for adults and children to take.

The usual dose for adults is five Celloids (tablets) of the indicated remedy. For children about one-half the quantity. The Celloids can be given dry on the tongue or dissolved in water, at intervals according to the severity of the case. When two or more remedies are used they should be taken in alternation. For instance, if *Ferrum Phos.* and *Kali Mur.* are to be used in alternation every hour, and the first dose of *Ferrum Phos.* is taken at 8 o'clock A. M., then *Kali Mur.* should be taken at 9 o'clock and *Ferrum Phos.* again at 10 o'clock, etc.

In very acute cases of severe pain, cramps, etc., the remedies should be taken frequently—about 10 minutes apart, in hot water, if possible. Less frequently after relief is obtained.

In chronic cases, from three to four doses per day.

It is sometimes found necessary or advisable to give what is termed an "*intercurrent*" remedy. This consists of a single dose of this remedy either once, twice or three times per day (generally morning and evening), in connection with the main remedies.

All of the biochemic remedies will work better and quicker in hot water, in severe cases. Further directions for giving the remedies will be found fully noted under the various diseases.

EXERCISE

This is of the utmost importance, and should *never* be neglected.

The method of exercise should be selected according to the necessities of each individual. Walking is excellent, but only the lower muscles are brought into use. Horseback riding, if the horse be not too "rough," is beneficial. Golf and tennis have come especially into prominence as healthful exercises, for not only are many muscles called into action, but the mind is constantly occupied. Rowing is an exercise of the same character, but is unavailable to many. Driving is a very poor method of exercising, but is better than nothing.

All exercise should aim to occupy the mind as well as the body. To forget one's self for the time is often a pleasure and will reflect its beneficent influence on the body. To be compelled to take a walk or ride is work, and the benefit derived will not be so great as were the exercise anticipated and enjoyed. Exercise should be daily—should be within proper bounds—should be adapted to the individual, and, above all, should be attended with pleasure.

THE BOWELS

The bowels should be kept as regular as possible. The colon is the sewer of the body; if it becomes clogged, the toxic matter is absorbed into the system and some form of disease usually results. One movement each day is the average, although there may be extreme cases, either way, without apparent impairment of the general health. Regularity of the habit is an absolute necessity to good health. This fact is almost universally believed, and medicine venders, never slow to take advantage of an opportunity, have flooded the market with every conceivable compound, capsule and pill to act upon the alimentary canal. The practice of taking these drugs indiscriminately is equally pernicious in its effects as habitual constipation.

Many of the preparations now before the public leave the bowels in a much worse condition than before using. If anything should be needed besides proper exercise, the author would recommend the use of a simple mild laxative remedy.

Massaging the abdomen, eating coarse foods, bran, apples, etc., are very beneficial. The patient should avoid straining while at stool, as it is likely to cause piles, uterine prolapsus, etc. Enemas of hot water in which a little common salt has been dissolved is of great benefit, if enough is taken to thoroughly flush the colon. In fevers and other low diseases, an occasional flushing of the bowels is of the utmost importance. It will lower the temperature and soothe the patient much better than any cathartic will do. If I were to sum up, in two words, the best methods of keeping the bowels in good condition, I should say, *exercise* and *regularity.* The rigid practice of these would soon have a marked effect upon our general health.

BLADDER

Trouble in the bladder, as evidenced by abnormality of the quantity or quality of the urine, with or without pain during urination, is often met with in people of all ages, from babies up to those of extreme age.

It frequently arises from a cold, overstrain at work or play, falls, the overuse of condiments, such as pepper and other spices, too much alcohol, or anything which tends to overstimulate the bladder.

Occasionally it may be caused by disease of the kidneys, and when the bladder trouble does not yield readily to treatment, kidney disease should always be suspected, and a search for kidney symptoms made, as well as an examination by a physician where possible.

The fact that an acute bladder trouble may also become chronic if neglected, should induce the sufferer from frequent bladder disorders to have a urinary analysis made by a competent physician, so that the real cause of his illness may be found and removed. This should also be done if blood appears persistently in the urine, or if the color of the urine remains constantly changed from normal. Sharp and continued pain along the urethra or urinary passage is another warning.

DIET

The diet, both in health and during disease, must be suited to the peculiarities of constitution in individuals. That which is food for one, sometimes is poison to another. For instance: while milk is a generally healthful food, some persons cannot take the smallest quantity without serious inconvenience; others throw out a rash after partaking of fish; and others, still, cannot bear certain kinds of fruit. Each individual must be a law unto himself.

Of course, during the existence of disease, certain kinds of food should be prohibited, but as a rule, under biochemic treatment, the practice of eating reasonable quantities, well masticated, is all that is necessary. Under this rule, if rigidly enforced, the diet may consist of almost anything within reason. In fevers, and other highly inflammatory conditions, meats and foods which are difficult to digest should be avoided.

The appetite is Nature's method of choosing that which the system needs, but it should always be governed by reason, not so much in the selection of foods, but in regard to their quantity and preparation. A distinction must be drawn between the *true* and the *false* appetite. The latter is generally acquired, and craves spirituous liquors, highly-seasoned food, etc., which is detrimental to health.

During the prevalence of disease the appetite is a true indicator of the needs of the system. It will frequently call for things which, to our preconceived ideas, seem outrageous, but Nature knows best, and a careful observance of her wishes will bring grand results. But in obeying the "cries" of Nature, we should, as I said before, be cautious as to the quality and quantity given. In chronic cases the rule to follow is: eat those things which seem to agree, and *avoid* those which disagree.

For beverages, water, toast-water, milk, milk and water, cocoa, unspiced chocolate, arrow-root, gruel, barley-water, sugar and water, rice-water, and, in some cases, weak black tea—is sufficient for all purposes.

During the course of an illness it is advisable to avoid green tea or strong black tea, coffee, malt liquors, wine, spirits and stimulants of every description.

HABITS

To quote a celebrated authority on the subject: "As regards habits, it may be briefly observed, that a regular method of living, avoiding ill-ventilated apartments, late hours, dissipation, overstudy, anxiety, and other mental emotions, and taking sufficient air and exercise, are the best preservatives of health."

As the above covers the whole subject, nothing more need be added.

THE GENUINE BIOCHEMIC REMEDIES

Great care should be exercised in purchasing the Schuessler Biochemic remedies. Some dealers have resorted to substitution, therefore it becomes very necessary that great care should be taken, and no remedies purchased except from a pharmaceutical house which is in every way reliable.

THE PREVENTION OF SEASONAL ILLNESS

During the Winter, coughs, colds, pneumonia, bronchitis, and similar troubles of the nose, throat and lungs are prevalent. These are caused by a deficiency of *Kali Mur.* and *Ferrum Phos.* in the system, brought about by the sudden changes of temperature and exposure to cold, drafts of air, etc. By taking five celloids of *Ferrum Phos.* in the morning, and five celloids of *Kali Mur.* in the evening, and thus maintaining a reserve supply of these salts for the body to call upon at need, disease may frequently be prevented, and an attack of possibly dangerous illness warded off.

In the Spring a condition of anemia exists in the majority of people, manifested by fatigue, lassitude, inability to concentrate the mind on work, and a vague feeling of discomfort, frequently known as "spring fever." *Ferrum Phos.* and *Calcarea Phos.* are the lacking salts which cause anemia. The persistent use of five celloids of *Ferrum Phos.* in the morning, and five celloids of *Calcarea Phos.* in the evening, during the months of March, April and May, is a wise defense against the possibilities of "spring fever."

During the Summer months, digestive and intestinal troubles are most frequent, and to provide a reserve supply of cell-salts, five celloids of *Kali Mur.* in the morning and five celloids of *Natrum Phos.* in the evening, should be taken regularly during June, July and August.

The Fall months are the period when the sudden changes of temperature and chilly nights act powerfully on systems weakened by heat of Summer, and as a protective measure, five celloids of *Ferrum Phos.* in the morning, and five celloids of *Kali Mur.* in the evening, should again be taken.

In this way, throughout the year, a reasonable amount of precaution may be taken against disease. Furthermore, during epidemics of "flu," hay fever, measles, and other troubles, the cell-salts lacking in such diseases, which may be found by consulting the corresponding article in this book, should be taken regularly as a preventive.

"An ounce of prevention is worth a pound of cure."

The Twelve Biochemic Remedies

How They Act on the System—Their Chief Uses

CALCAREA FLUOR.

Calcarea Fluor. works with albumen to make elastic fiber. It is a constituent of the enamel of teeth, connective tissue, and the elastic fiber of all muscular tissue.

A deficiency of elastic fiber in muscular tissue causes a relaxation of the tissue, and is a primary condition in a large number of diseases.

It is indicated in all ailments which can be traced to a relaxed condition of the elastic fibers, including dilatation of the blood vessels, hemorrhoids, enlarged and varicose veins, hardened glands.

It is also indicated in disease affecting the covering of bones and the enamel of the teeth.

The symptoms of this remedy are all worse in damp weather, and are relieved by rubbing and fomentations.

CALCAREA PHOS.

Calcarea Phos. is a constituent of the bones, teeth, connective tissue, blood corpuscles, and the gastric juice. It unites with the organic substance albumen, giving solidity to the bones, building the teeth, and entering into all the important secretions of the body, such as the blood and gastric juice.

Bone is fifty-seven (57) per cent *Calcarea Phos.,* and without it no bone can be formed.

Calcarea Phos. uses albumen as a cement to build up bone structure. It enters largely into the formation of teeth, hence is a valuable remedy in childhood.

Calcarea Phos. is also found in the gastric juice, and plays an important part in assimilation and digestion. It is closely allied, in some respects, to *Magnesia Phos.,* and is frequently given in alternation with this remedy.

The sphere of *Calcarea Phos.* includes all bone diseases, whether inherited or due to defective nutrition. It is the remedy in anemia and chlorosis; convulsions and spasms in weak scrofulous subjects; in teething, when the teeth are slow to make their appearance or decay too rapidly. In convalescence after acute disease, and in chronic wasting diseases, it acts as a tonic, building up new blood corpuscles.

Deficient development of children and young people; emaciation. It aids the union of fractured bones. Cold, motion, change of weather, and getting wet, generally aggravates the symptoms. Relieved by rest, warmth, and by lying down.

CALCAREA SULPH.

Calcarea Sulph. is found in the epithelial (or skin) cells and in the blood, and acts as a preventive of cell disintegration and suppuration.

A deficiency of this salt allows suppuration to continue too long.

It is indicated in the third stage of all suppurative processes, including catarrhs, boils, carbuncles, ulcers, abscesses, etc. It is also indicated in pimples and pustules of the face.

Silicea hastens the suppurative process, while *Calcarea Sulph.* closes up the process when the proper time comes.

All suppurations do not call for this salt, except in connection with some other, but the true indication is a thick, heavy, yellow pus or matter, and sometimes mixed or streaked with blood.

Calcarea Sulph. symptoms are aggravated by getting wet or by washing or working in water.

FERRUM PHOS.

Ferrum Phos. is the great remedy for inflammatory conditions. It is found in the blood, where it colors the corpuscles red and carries oxygen to all parts of the body. It gives strength or toughness to the circular walls of the blood vessels, especially the arteries.

Without a proper balance of *Ferrum Phos.* in the blood, health cannot be maintained. It colors the blood corpuscles red. A deficiency of this salt is the cause of all inflammatory conditions, colds, coughs, etc.

For all such conditions, whenever there is inflammation, under whatever name it may be known, *Ferrum Phos.* is the chief remedy.

It is indicated in all cases depending upon a relaxed condition of the muscular tissue, and in abnormal conditions of the corpuscles of the blood themselves. In all febrile disturbances and inflammations, at the commencement, before exudation has begun. The symptoms of these disturbances are: flushed face, fever, quick, full pulse, hot, dry skin, thirst, pain and redness of the parts.

In anemia it is excellent for its tonic action.

Ferrum Phos. symptoms are always aggravated by motion and relieved by cold.

KALI MUR.

Kali Mur. unites with albumen, forming fibrin, which is found in every tissue of the body, with the exception of the bones. A deficiency of this salt with a consequent release of albumen, causes a discharge or exudation of a thick, white, sticky character from the mucous membranes, and a white or gray coating of the tongue.

The chief indication for this remedy is the white or gray exudation, coating of the tongue or mucous lining of throat or tonsils. It is indicated in glandular swellings, discharges or expectoration of a thick, white, fibrinous consistency, white or gray exudations.

Excellent in catarrhal conditions with the above symptoms. It is the chief remedy in spasmodic croup, diarrhea and bronchitis, to control plastic exudation. It should be given in alternation with *Ferrum Phos.* in coughs, catarrh of the Eustachian tubes, skin eruptions with small vesicles containing whitish-yellow contents, and ulcerations with swelling and white exudations. Leucorrhea with above colored discharges.

Symptoms are generally worse from motion; stomach and abdominal symptoms are aggravated after taking rich and fatty foods.

KALI PHOS.

This is the great nerve and brain remedy. *Kali Phos.* unites with albumen and other organic matter to form the gray matter of the brain.

When nervous symptoms arise, it is due to the fact that *Kali Phos.* molecules have been overdrawn and a deficiency occurs. *Kali Phos.* is the only true remedy for this condition, because nothing else can possibly supply the deficiency.

Wherever a disease can be traced to a nerve degeneracy, we enter the field of *Kali Phos.*

Some of its indications are as follows: Loss of mental vigor; poor memory; prostration; depression; nervousness; neurasthenia; brain-fag; offensive breath; foul diarrhea or dysentery; dizziness and vertigo due to nervous exhaustion; incontinence of urine from paralysis of sphincter muscle. Tongue coated as if spread with dark, liquid mustard.

Many of its symptoms are aggravated by noise; by physical or mental exertion; by beginning to move after rest; pains worse in cold air. Symptoms are relieved by gentle motion, eating, rest, excitement or anything which diverts the mind and aids in restoring the deficient nerve force.

For its exact therapeutic application, see under the different diseases.

KALI SULPH.

Kali Sulph. is a carrier of oxygen, as well as of organic material, to the cells of the skin. It furnishes vitality to the epithelial tissues, and is a constituent of the scalp.

It has an affinity for oil, hence its secretions are sticky, slimy, etc. It corresponds to the third stage of all inflammatory conditions, when the secretions are light yellow, slimy, sticky, watery or greenish matter.

Slimy, yellow coating on the tongue. When the exudation from the respiratory organs is slimy, thin, yellow and watery.

Skin diseases, with sticky yellowish secretions, yellow scales.

Dandruff, yellow scales; catarrh of the stomach, with slimy, yellow coating on the tongue; catarrh of bowels, leucorrhea, diarrhea, etc., if the discharges answer the above description. Menstruation, too late and scanty, with weight and fullness in abdomen; sudden suppression or retrocession of eruptions of measles, smallpox, scarlet fever, etc. It promotes perspiration, opens the pores of the skin, and throws the blood to the surface.

All of its symptoms are aggravated in a warm room and toward evening, and are relieved in the cool, fresh air.

MAGNESIA PHOS.

The work of *Magnesia Phos.* is chiefly confined to the delicate white nerve-fibers of the nerves and muscles.

It uses albumen and water to form the transparent fluid which nourishes these white threads or fibers.

A deficiency of this salt in the fiber allows it to contract, hence it produces spasms, cramps, convulsions, etc. When this contraction takes place, there is pressure on the sensory nerves, and this gives rise to sharp, shooting, darting or neuralgic pains in any part of the body.

This salt is Nature's *anti-spasmodic;* and given in hot water will produce splendid results.

In spasmodic conditions, the end to be desired is to relax the muscles, and can be accomplished only by relaxing the motor nerves.

This can be done, easily and naturally, by supplying the deficiency of the cell-salt which has caused the trouble. *Magnesia Phos.* is indicated in all diseases having their origin in the white nerve-fibers. It is particularly indicated in lean, thin, emaciated persons of a highly nervous temperament. It should be given in all forms of spasms, in cramps, St. Vitus' dance, spasmodic retention of the urine, colic, palsy, painful menstruation.

Neuralgic pains (sharp, shooting, darting) in the head, face, teeth, stomach or abdomen, call for this remedy. The patient is languid, tired and easily exhausted.

All pains are lightning-like, shooting or boring, and change their location frequently.

Magnesia Phos., pains are worse on the right side, from cold air, cold water and by touch. They are relieved by heat, firm pressure, friction, and by bending double.

In acute diseases, give *Magnesia Phos.* in hot water, as heat aids the action of the remedy.

NATRUM MUR.

Natrum Mur. works with water and properly *distributes* it through the organism. The body contains more of this cell-salt than any other, except *Calcarea*

Phos. The necessity of this will be seen when we learn that our bodies are composed of about seventy (70) per cent water, which in the absence of *Natrum Mur.* would be inert and useless. It is the power that this cell-salt has to use water that renders it of value to mankind.

The same principle holds good in vegetable life.

Any deficiency of this cell-salt causes a disturbance of the water in the human organism, because it has lost that element which renders it fit to perform its allotted task.

There sometimes occurs an excessive dryness of some mucous membrane, while another may be discharging copiously a watery fluid. This is due to an unequalization of the water in the system, and *Natrum Mur.* is the proper cell-salt to restore the equilibrium. It acts on the blood, liver, spleen, and every mucous membrane of the body.

Natrum Mur. is indicated in headache, toothache, faceache, stomachache, etc., when there is either salivation or excessive secretion of tears, or vomiting of water and mucus. Also catarrhal affections of mucous membranes; head colds, with secretion of transparent, frothy, watery mucus. Small watery blisters or blebs on the skin; diarrhea, slimy, transparent stools; inflammation of the eyes, with discharge of tears; leucorrhea, watery, smarting or clear, starch-like discharge.

Tongue clear, slimy, small bubbles of frothy saliva on the edges, sometimes salty taste in the mouth. Pains are periodical, worse in the evening.

NATRUM PHOS.

Natrum Phos. is the remedy in all cases where there exists an excess of acid in the system.

Natrum Phos. splits up lactic acid into carbonic acid and water, and throws it off through the lungs. It has an affinity for sugar, and assists in eliminating any excess from the blood. A lack of proper balance of the alkaline cell-salt in gastric juice will allow ferments to arise, and so retard digestion that the lining of the stomach quickly becomes involved. *Natrum Phos.*, by its affinity for lactic acid, is indicated in diseases wherein the acid is apparently in excess. It relieves sour belchings and risings of fluids; sour vomiting; greenish, sour-smelling diarrhea; colic, spasms and fever, when due to acid conditions of the stomach; all gastric derangements, when acidity is present; intestinal, long or pin worms; yellow, creamy discharge from the eyes.

The tongue indication for the use of *Natrum Phos.* is a moist, thick, golden-yellow coating, either on the tongue or palate.

All exudations which are creamy, golden-yellow call for this cell-salt.

NATRUM SULPH.

This cell-salt is found in the intercellular fluids, and its principal office is to regulate the *quantity* of water in the tissues, blood and fluids of the body. It has an affinity for water to that extent that it eliminates the excess from the blood and blood serum. It also works with the bile, and keeps it in a normal consistency.

Natrum Sulph. splits up lactic acid into carbonic acid

and water: this leaves a residue of water to be gotten rid of.

Natrum Sulph. performs this work; each molecule of the cell-salt has power to take up and carry out of the organism two molecules of water.

Remember the action of these three sodium salts. *Natrum Phos.* creates water, by breaking up lactic acid; *Natrum Mur.* distributes water; and *Natrum Sulph.* regulates the quantity of water in the system.

Natrum Sulph. is indicated when there is a dirty, brownish-green, or grayish-green coating on the root of the tongue; dark green stools from excess of bile; jaundice; biliousness; excess of bile, bitter taste, greenish diarrhea and vomiting of bile. Bilious headache. Intermittent fever with vomiting of bile. Erysipelas, with smooth, red, shiny skin, also swelling of the skin. Vomiting in pregnancy, with bitter taste; watery yellowish secretions on the skin.

Natrum Sulph. symptoms are worse in the morning and in damp, rainy weather; better in dry, warm atmosphere.

Symptoms aggravated by using water in any form. Living in low, marshy places, damp buildings, basements, or eating water plants, fish, etc., will cause a molecular disturbance in *Natrum Sulph.* and produce diseases similar to the above.

SILICEA

Pure Silica is found in the hair, nails, skin, covering of bones and nerves. It promotes the discharge of pus and should therefore be used when there is a hard lump in the flesh, with suppuration or threatened suppuration.

Silicea is indicated in nasal catarrh, with fetid, offensive discharge; tonsillitis after pus has begun to form; thick, copious expectoration, profuse night sweats and great debility; carbuncles, boils and ulcers, if deep-seated and discharging thick, heavy, yellow pus; gout.

The Application
of the
Twelve Schuessler Biochemic Remedies

APHTHAE
(Canker—Thrush)

CAUSES AND SYMPTOMS

Aphthae is a disturbance which occurs most frequently in little children although adults are not immune to it. It arises from a disordered stomach, the symptoms of which usually point to a deficiency of the cell-salt *Kali Mur.* A contributory cause is the neglect of sanitary precautions, such as cleansing the sterilizing bottles, nipples, etc., used in nursing the child.

Canker sores appear on the membranes of the mouth, usually in the form of small, round whitish vesicles, which sometimes spread into small ulcerated patches.

TREATMENT WITH THE SCHUESSLER REMEDIES

Kali Mur.—Is the chief and, generally, the only remedy required.

Natrum Mur.—If there is much "drooling" of saliva.

Ferrum Phos.—If fever symptoms are present.

Dose for children: Three celloids every one to two hours.

SUGGESTIONS

Carefully swab and cleanse the mouth with a solution of Creozone or Boric Acid. Temporarily omit all food which is not easily tolerated by the child.

AMENORRHEA

(Absence of Menstruation)

CAUSES AND SYMPTOMS

Amenorrhea is a failure of occurrence of the menstrual flow. It is called "Primary Amenorrhea" when menstruation has never occurred—and "Secondary Amenorrhea" when menstruation has been established and later suppressed.

Primary amenorrhea is frequently due to faulty or delayed development of the female generative organs or to an illness which causes low general vitality.

Secondary amenorrhea may be caused by pregnancy, menopause (change of life), or by illnesses such as anemia, consumption, nervous debility, etc., in which mineral salt deficiency is a predominant factor.

TREATMENT WITH THE SCHUESSLER REMEDIES

Calcarea Phos.—When the suppression arises from anemia or from faults of diet. Gradual suppression, pale face, tired, languid, no ambition.

Kali Phos.—Suppression due to mental strain, exhausting occupation, etc. Depression of spirits, lassitude and general nervous debility.

Kali Mur.—Suppression due to taking cold, wet feet, etc. White coated tongue and general inactivity of the glandular system.

Kali Sulph.—Menses too late or scanty, indigestion, tongue yellow and slimy.

Dose: Five celloids of the indicated remedy four times daily. If more than one remedy is required, they should be taken in alternation.

SUGGESTIONS

Other remedies may be found necessary in alternation with the remedy indicated for the suppression. For spasms, convulsions, etc., *Magnesia Phos.*; rush of blood to the head or chest, cold feet, etc., *Ferrum Phos.* In such cases the diet should be very light, but in anemic conditions without any severe symptoms the diet should be plain and nourishing. An occasional hot sitz-bath is beneficial to cause an engorgement of blood to the parts where needed.

Change of air, scenery, occupation, etc., is greatly to be desired, and there should be plenty of outdoor exercise without incurring fatigue.

CLINICAL REPORT

Miss H., age 16, high school pupil. Menstruates very irregularly, once in three or four months. No menses for several months past. Nervous and sensitive. Slightly jaundiced. Dark streaks under eyes. Skin dry and rather inclined to be scaly. Poor appetite. Is losing flesh. Gave *Kali Phos.*, a dose three times a day. Two months later the young lady's mother reported that Miss H. was menstruating regularly and was much improved in general health.

(L. C. S., M.D.)

ANEMIA

CAUSES AND SYMPTOMS

Anemia is a lowering of the quality of the blood. The blood-cells may be fewer in number or the oxygen-carrying red constituents of the blood-cells may be lessened.

Anemia may result directly from a loss of blood or from a wasting disease attended with a deficiency of the vital cell-salts needed to build new blood-corpuscles. A lack of *Calcarea Phos.* is responsible for the loss of white blood-cells and a *Ferrum Phos.* deficiency is responsible for the lowering of hemoglobin (red blood).

Anemia can be easily recognized by the characteristic look of bloodlessness, pale or greenish-white face, wounds heal slowly and in women it frequently causes irregularities of menstruation.

TREATMENT WITH THE SCHUESSLER REMEDIES

Calcarea Phos.—Is the chief remedy in this disease, and should be given in alternation with other indicated remedies. Face pale, greenish-white, bloodless. Anemia when nutrition is deficient; infants are thin, delicate and puny. Anemia after wasting or exhausting diseases. An important remedy in chlorosis.

Ferrum Phos.—An important remedy when there is deficiency of red blood-cells. Face pale but flushes easily; lips pale, headaches; chlorosis.

Natrum Mur.—Anemia in young girls, wasting discharges from mucous membranes. Leucorrhea in place of menstruation. Skin dry and scaly, constipation with dry stools. In alternation with *Calcarea Phos.*

Kali Phos.—Anemia from long-continued mental strains, causing depression of the mind. After worry.

Kali Mur.—When eruptions of the skin exist in connection with anemia. In alternation with *Calcarea Phos.*

Natrum Phos.—Anemia in connection with indigestion, acid risings, golden-yellow coated tongue, etc.

Silicea.—Anemia in infants when improperly nourished, or when scrofula is present. Intercurrently with *Calcarea Phos.*

Dose: Five celloids of the indicated remedy every two hours. When required two or more remedies may be used in alternation. The remedies, necessarily, should be taken regularly over a period of several months.

SUGGESTIONS

A proper diet is an important part of the treatment, it should be nutritious but must not cause disturbances in digestion. Red meats, liver, cream, eggs, green vegetables, are articles of food which should be eaten freely.

If the patient suffers from nervousness, any known causes of worry should be removed, if possible. Moderate exercise, out-of-door life, reasonable exposure to direct sunlight are measures beneficial to anemic persons.

So-called pernicious anemia is a serious disease which does not yield to simple treatment, and the patient should be placed under the care of a competent physician.

CLINICAL REPORTS

Harriet N., age 20, pale face, sort of a wax appearance. Sort of dull and listless, no appetite. Menses bad, irregular and scant.

Treatment: Calcarea Phos., Ferrum Phos. Six weeks later no further trouble.

(L. T. K., M.D.)

A girl about 16 years of age. It was not my case, but I was called to see her after her attending physician had given her up and the priest had prepared her for death. The case seemed hopeless. Her heart was greatly enlarged, hemorrhages were frequent, she could not lie down. She had been dosed with iron in various forms and the stomach would not tolerate the crude doses. It was evident that the strong medication had defeated the doctor's effort in-as-much as they were so crude that the stomach would not take them, and, of course, the system would not assimilate them. She was literally starving to death for the want of iron and phosphates.

I prescribed *Ferrum Phos.* and *Calcarea Phos.* in celloids, to alternate them hourly during the night, and if she lived I would see her in the morning. The morning found her slightly improved. She had kept the medicine down, there had been no bleeding, and she had rested. The case was clear. Her system took the potentized remedies as a sponge takes water. She began to mend. In a few weeks she had recovered and later she married. The last I knew of her she was well and had a family of children.

(E. M. R., M.D.)

APPETITE

A deranged appetite is usually indicative of some disease of the system. Frequently a morbid appetite is simply a call of Nature to give the system a rest, but if it continues any length of time a serious diseased condition may exist, and should receive proper attention. Sometimes it becomes quite a factor in the progress of continued diseases, but wherever found it should receive the same treatment. It is one of the signs which Nature throws out to guide the practitioner in a correct diagnosis. If there is a repugnance for any special or all food, it is a pretty sure indication that Nature does not want it. It sometimes becomes necessary to "tempt" the appetite, not for the purpose of forcing on Nature that which she does not wish, but to find that food which is palatable and acceptable to the system.

TREATMENT WITH THE SCHUESSLER REMEDIES

Ferrum Phos.—Loss of appetite, with feverish conditions.

Calcarea Phos.—Loss of appetite, when indigestion or poor assimilation is present. An excellent remedy to promote digestion, especially in anemic conditions.

Natrum Phos.—Loss of appetite with acid symptoms of the stomach, "Heartburn," etc.

Kali Phos.—Voracious appetite after typhoid fever or other wasting diseases, indicating a poor state of the blood. Hungry feeling after taking food. "Gone feeling" in the pit of the stomach.

SUGGESTIONS

Never force any particular food upon a patient; tempt the appetite if possible. More harm than good is done by compelling a patient to partake of food repugnant to him or her.

The belief, however, that certain foods or combinations of food are injurious is very frequently not based upon facts. There are, it is true, persons who show a genuine intolerance (called allergy) for certain foods. Even small quantities of these foods may cause disturbances such as skin eruptions, headaches, acidity, colic, etc. Advice: Avoid all food which does not agree, and do not overload the stomach with food you like.

APOPLEXY

The most valuable achievements in medical science will be attained in the field of disease prevention.

The destructive or damaging effects upon the tissues and organs of the body caused by certain serious diseases are frequently very difficult to correct and sometimes cannot be repaired by any known method of treatment.

The majority of cases of apoplexy are caused by the rupture of a blood vessel of the brain. A violent mental emotion, for instance, may be the immediate exciting cause of apoplexy, but the real underlying cause is an abnormal condition, probably brought about by a long existing deficiency of certain indispensable cell-salts. The timely correction of this deficiency would in many cases prevent a condition predisposing to apoplexy, but this may not prove effective in repairing the damage done by a ruptured blood vessel.

Apoplexy does not necessarily indicate the rupture of a blood vessel, but in any event it is a serious condition, which demands the services of a competent physician, and for this reason we only give a few suggestions for the emergency care of the patient until a physician arrives.

Put the patient to bed, if possible, in a cool room. Loosen the clothes, especially about the neck; the head should be higher than the body; feet and legs to hang over the edge of the bed. Place the feet into warm water, or rub them briskly. If the patient is able to take medicine give *Ferrum Phos.,* the remedy for congestion, bleeding, etc. Give five celloids every 15 minutes.

APPENDICITIS

Appendicitis is an inflammation of the appendix, which is at the point where the small bowel empties into the large one and located in the lower right corner of the abdomen. Appendicitis is so common and so dangerous a disease, so sudden in onset and so quickly reaching a serious stage, that it is here described in order that it may be recognized and the patient placed in the care of a physician soon enough that an operation, if needed, may be done before it is too late.

Appendicitis usually begins with a chill and vomiting followed by the onset of pain and fever. The pain is at first located about the navel, gradually spreading but finally centering in the area of the appendix. The fever is not high, very rarely going above 102 degrees. From the beginning of the disease an area of tenderness will be found over the location of the appendix, the abdominal wall at this point will be hard and rigid and the right leg usually drawn up to ease the strain over the sore spot.

The bowels are usually constipated, but after the inflammation begins it is important that the bowel be kept at rest in order to prevent the spread of the infection; therefore *no cathartic* should be given, but the lower bowel may be cleared out by giving an enema of warm soap-suds.

No food should be given the patient with acute appendicitis until the temperature is normal, showing that the inflammatory process has subsided. In severe cases the amount of water should be very limited and thirst can be allayed by giving small amounts by rectum.

As the loss of life from this disease is very great and the risk of operation mounts rapidly with the expiration of each period of twenty-four hours after the onset of symptoms, it is very important that the patient be seen by a physician as soon as possible after the beginning of the trouble. Therefore no treatment is given here for this disease.

ARTHRITIS

CAUSES AND SYMPTOMS

Arthritis is a disease affecting the structures in and about the joints. The causes apparently are the combination of a specific germ infection with a basic deficiency of some of the mineral salts which disturbs the normal process of metabolism, weakening the power of resistance of the system to infection.

The acute form of arthritis resembles rheumatism, the joints are red, inflamed, swollen and painful. The chronic form is characterized by a gradually increasing amount of solid deposits in the joints, thus limiting motion. This form frequently resists almost every method of treatment.

TREATMENT WITH THE SCHUESSLER REMEDIES

Ferrum Phos.—In acute attacks with fever, inflammation of the joint which is swollen and red, joint very painful, worse from motion.

Natrum Phos.—Acid conditions, acid stomach, bitter taste, golden-yellow coating at the root of the tongue.

Kali Mur.—Acute cases with swelling in the joint, but little fever. White coated tongue.

Natrum Mur.—Early stages of hard deposits in joints. Joints seem to be dry, creaking and rubbing sounds on motion. Constipation.

Calcarea Phos.—Pain in the joints, worse at night and when at rest. Affected parts feel cold, deep-seated pains.

Calcarea Fluor.—Stony hard enlargements on the joints, especially on the finger joints.

Dose: In acute cases—five celloids of the required remedies in alternation every hour. In chronic conditions—five celloids three or four times daily.

SUGGESTIONS

In acute cases the affected joint should be kept at rest, warm and protected from exposure. Dry heat applications will help to relieve pain. Gentle massage with an analgesic balm is also palliative. In chronic arthritis the patient seems to be more comfortable if he keeps in motion as much as possible, as prolonged rest seems to aggravate the stiffness of the joints.

ASTHMA

CAUSE AND SYMPTOMS

According to Dr. Schuessler a deficiency of certain mineral salts in the system plays an important part as a causative factor of asthma, and the supplying of the proper cell-salts is one of the first requisites toward correction of this disorder.

The most characteristic symptoms of asthma are the spasmodic attacks of oppressed, difficult, labored breathing, aggravated by lying down and by damp weather. The attacks may last a few hours and in some cases several days. It is usually a chronic condition with periods of acute attacks, sometimes associated with other ailments such as chronic bronchitis, nervousness, etc.

TREATMENT WITH THE SCHUESSLER REMEDIES

Kali Phos.—Is the chief remedy for the oppressed breathing; in large and frequent doses during attacks. Nervous asthma; asthma from taking the least food; nervous system depressed.

Kali Mur.—Asthma when derangements of the stomach are present; white coated tongue; costive bowels; sluggish hard liver. Expectoration is thick, white, tough mucus, hard to cough up. Alternate with *Kali Phos.* for the breathing.

Magnesia Phos.—Asthma with troublesome flatulence or constrictive sensation in the chest. Spasms of the bronchial muscles. Alternate with *Kali Phos.*

Natrum Mur.—Asthma with expectoration of clear, frothy mucus; watery discharges from the eyes and nose. Alternate with *Kali Phos.*

Calcarea Phos.—Intercurrently with the indicated remedies in all cases. Asthma in children; mucus clear and tough.

Calcarea Fluor.—When the expectoration consists of tiny yellow lumps of matter, raised with difficulty.

Natrum Sulph.—Asthma of young people, with bilious conditions; worse from damp weather or wet surroundings; greenish-yellow coating on the root of the tongue; expectoration greenish and very copious.

Dose: Five celloids of the indicated remedies in alternation every 15 to 30 minutes during the acute stage, less frequently after the attacks have subsided. During the intervals between attacks the required remedies should be taken three or four times daily.

SUGGESTIONS

In most cases the use of *Kali Phos.* will procure prompt and satisfactory relief, but if necessary a cup of strong black coffee may be taken. Inhaling of the fumes of burning nitre-paper will often prove efficacious as a relief-measure.

Asthmatic patients should avoid stimulating or irritating foods and drinks. Heavy meals at night should be avoided, also violent exercise. Patients who find their condition aggravated by damp air sometimes find it necessary to move to a dry climate in order to obtain complete or permanent relief.

CLINICAL REPORTS

In 1904 Mr. B., a confirmed asthmatic, called me to treat him during an attack. Tall, slim, stoop-shouldered. He was forced to sit up and lean over the back of a chair. Spasmodic cough, raising a tough white phlegm. I gave him *Kali Mur.* every 20 minutes until relieved of the attack and then every three hours.

This remedy completely cured this man. He came to my office more than 700 miles from his home four years ago and told me that he had never had a severe attack of asthma since.

(R. C. B., M.D.)

One winter night I was called to visit a woman suffering with asthma. She could not speak, being so terrifically choked up. I managed to get a dose of *Natrum Sulph.* in her mouth and she managed to get it down. I waited five minutes then asked if she was any better. She moved her fore-finger. I waited ten minutes longer and got another dose into her. In 15 minutes she was able to speak. I prepared some more of the same medicine in a tumbler half full of water and ordered it given in teaspoonful doses, one every half hour. She continued taking the medicine every two to four hours for three weeks.

Three years later I was called to treat her again with the bronchitis but it did not develop into asthma as it did before.

I can honestly say that this is the only case of asthma coming from that cause I have ever cured. In the following seven years I had known her she continued free from it.

(E. B. F., M.D.)

BACKACHE

CAUSES

Backache is often associated with muscular rheumatism or lumbago. It is also frequently caused by sprains of the muscles or of the spine from heavy lifting, jumping, etc. Occasionally backache arises from female disorders or kidney troubles, and in some cases it is simply an early symptom of influenza.

The restoration of a proper balance of the mineral cell-salts is necessary for a return to normal in cases not due entirely to physical injury.

TREATMENT WITH THE SCHUESSLER REMEDIES

Calcarea Fluor.—Dragging pain in the lower part of the back, particularly when associated with displacement of the female organs, or with hemorrhoids.

Ferrum Phos.—Acute lumbago, pain in muscles, especially when in motion.

Magnesia Phos.—Neuralgic pains along the spine. Pains are sharp and shooting, relieved by heat but not by rest.

Kali Phos.—Lameness in back, aggravated by motion. General condition of nervousness.

Natrum Sulph.—Rheumatic pains in back, worse in damp weather and at night.

Dose: Five celloids of the required remedy (when more than one remedy is required take them in alternation) every hour during the acute stage, less frequently after relief is obtained.

SUGGESTIONS

Nearly all forms of backache can be relieved by rest and by applications of heat. Gentle massage with an analgesic balm, is helpful, as it stimulates circulation. Irregularities of bowel movements should be corrected.

BOILS

(Abscess—Carbuncle)

CAUSES AND SYMPTOMS

Boils, carbuncles and abscesses are local inflammations, terminating in a collection of pus in the tissues of the body, and are produced by infection from certain germs. The principal underlying and contributing cause, however, according to Dr. Schuessler, is a deficiency of certain of the mineral salts, which are always present in health. Such deficiencies cause waste material in the tissues, forming a medium in which germs can grow and multiply.

In the early or acute stage of an abscess there is heat, inflammation, swelling and pain in the affected part, which is usually followed by the formation of pus, and finally by a discharge of the waste material.

Abscesses may appear on various parts of the body, such as in the tonsils, the gums near the roots of the teeth and frequently on the breast in nursing mothers. The most common type of abscesses are known as boils, which may appear on any part of the body, but are generally located at the base of the hair-follicles, through which the germs invade the tissues. Carbuncles are similar to boils, but more deep-seated and larger; they have more than one opening on the surface, and are more serious than ordinary boils.

TREATMENT WITH THE SCHUESSLER REMEDIES

Ferrum Phos.—First or inflammatory stage, when there is heat, pain, congestion and fever. Given early in alternation with *Kali Mur.* it will often prevent swelling and suppuration (formation of pus).

Kali Mur.—For the swelling, before pus has begun to form (alternate with *Ferrum Phos.*)

Silicea.—After *Kali Mur.,* when the swelling becomes soft and pus has commenced to form. It will assist suppuration, cause the abscess to ripen and often break without surgical interference. Should also be given after the boil has broken and is discharging its contents.

Calcarea Sulph.—After *Silicea,* if the discharge continues too long and the wound refuses to heal, owing to a torpidity of the tissues.

Calcarea Fluor.—When the suppurative process affects the bone. When the boil, abscess or suppurating wound has a hard, callous edge.

Natrum Sulph.—Fistulous abscess of long standing. Discharge of watery pus, and when the wound is surrounded by a dark, bluish border.

Calcarea Phos.—Intercurrently, to promote the formation of purer and "richer" blood. Also five celloids three times daily for several weeks following the healing of the boil or abscess, to prevent recurrence of the boil or abscess.

Dose: The required remedy should be taken in doses of five celloids, every hour in the early stage, and in chronic conditions three or four times daily. When two or more remedies are needed, take them in alternation.

SUGGESTIONS

In addition to the use of the proper Schuessler Remedies, local applications may be used with beneficial effects. In the early stage hot, dry applications to the parts are helpful in preventing suppuration. After pus has accumulated a hot wet dressing, or a drawing salve will hasten the suppurative process and promote normal healing. After the abscess has opened, the parts should be kept clean. Apply a dressing (gauze or cotton pad) moistened with an antiseptic solution (Creozone) until the pus stops draining. Keeping the surface moist will prevent a crust formation and the shutting off the drainage.

In some cases it may become necessary to lance the abscess to relieve the painful pressure from pus accumulations.

CLINICAL REPORTS

Mr. C., age 42 years, came to my office about first of April suffering with boils. He then had eight or ten boils of all sizes and stages and scars all over where they had gotten well. The patient said he had a crop of those boils each fall and each spring of the year and they had been repeating regularly for the past eight years. He had consulted quite a few doctors and had taken all kinds of blood medicine that was advertised on the market until he was wholly disgusted with it all. But he would give all he had if he could get a cure.

I told him I could help him if he would co-operate with me and take the medicine correctly. So he agreed to do as directed. In order to hasten those boils that were started, I prescribed *Silicea,* five celloids four times a day at 6 A. M., 10 A. M., 2 P. M. and 6 P. M. I gave him a small box of these celloids and also gave him *Calcarea Phos.* celloids, five celloids four times a day, at 8 A. M., 12 M., 4 P. M. and 8 P. M. I also instructed him to keep the bowels in regular order and to report when he was out of the medicine I gave him.

I saw him in three or four weeks with head up and feeling pretty good. He did not have a boil, but had a few small pimples, so I gave him more of the medicine and sent him away again.

That has been more than five years ago and he has not had a boil since.

(T. W. L., M.D.)

Here is a case of boils on the neck, treated with *Silicea.* Mr. E., a lawyer of this city, has been suffering for some time, and had not been relieved by lancing. He came to me—on examination I found two abscesses, molars. A new boil was forming, and his neck was too sore to do any extracting. I put him on treatment of *Silicea* for five days with great relief. The following week he had both molars extracted. I gave him another course of *Silicea,* and the boils have not returned.

(L. O. H., M.D.)

Mr. B., age 26. Four abscesses had formed and had been surgically treated. Never healed. Locations—right groin, inside of right knee, just above knee and right ankle. Long continued suppuration, thick purulent secretion. In bed for ten weeks.

Calcarea Sulph. selected on one symptom present,

pus with a vent. Dose, five celloids every four hours, four doses daily. Cured in five months.

(L. G. W., M.D.)

In February and March, 1929, while traveling in Germany, I had small hard lumps all over the right axilla extending finally to the breast, going on to suppuration (very angry looking).

In Hanover my relatives insisted upon my going to a hospital. I called and was informed an operation was imperative. I decided no unless unavoidable until I reached home and could enter the Nyack Hospital. (My son Frederick is on the staff (med.), a graduate of Columbia College of P. & S.).

Stayed in Hamburg three weeks waiting for our steamer and commenced taking *Calcarea Sulph.* and *Silicea* alternately every three hours, and before going on the ship was almost cured.

(M. E. O. S.)

Mrs. S., age 40. Past history: Much under weight for last 10 years. Very nervous. Has great many headaches, depressed. Periods irregular.

(R. L. E., M.D.)

A few years ago "acidity" was considered as being a counterpart of practically all diseases, and pharmaceutical houses stressed the alkaloidal properties of their remedies. But within recent months hyperalkalinity has come to be recognized as existing in numerous disorders, and the most outstanding example of this has apparently been brought to light in cancer research, where we are told that a high degree of hyperalkalinity has often been found to exist, making it necessary for the examiner to perform exhaustive urine, feces and blood tests to determine just in which direction acidity or alkalinity balances.

However, the writer's own case did not involve cancer. It had to do with almost constant headaches, extreme nervousness, furunculosis (frequently recurring boils), and a general feeling of weakness. Alkaloidal remedies, empirically administered, offered no relief. My brother wrote me of his results with *Silicea* in similar states, and so I determined to submit myself to a more searching examination than I had so far, an examination that we do not hesitate to give to our patients but which we may not be inclined to give ourselves. Considerable indican was found in the feces, and in the urine at the end of the day. The urine was found definitely alkaline, having some casts after an attack of headache, and being light colored and excessive in quantity.

I went on *Silicea* celloids, five celloids six times a day. Small boils that had persisted inside the nares rapidly cleared up. Headaches became less frequent, and the quantity of urine rapidly decreased until it became normal in amount and its color deepened. In three weeks' time much improvement was evident, and I now feel great general gain in feeling of well-being. I am rather tall and "lean," with large bony but small fleshy development, and I feel that people of that type tend to the accumulation of minerals in the system and that they may need more acid than they ordinarily obtain in foods. Therefore, I believe that *Silicea* (Silicic Acid) should be carefully considered by any physician who is administering the cell-salts of Schuessler. I feel that these remedies, being drugless, but consisting only of the elements contained in foods and in the body, can be administered profitably in any ailment of mankind, either separately or in conjunction with other indicated remedies.

(E. W. C., M.D.)

BILIOUSNESS

CAUSES AND SYMPTOMS

Biliousness is a term used to cover conditions arising from disturbances of the function of the liver associated with mineral salt deficiencies. There may be an excess production of bile, when the tongue will be coated dirty yellow or brown, the taste bitter, vomiting of bile, and headaches.

In other cases there may be lessened amount of bile with congestion of the liver, with inflammation of the liver and bile ducts. Here the tongue will be white, the stools constipated and light colored, and the bowels distended with gas. In severe cases jaundice may develop due to absorption of bile by the blood.

TREATMENT WITH THE SCHUESSLER REMEDIES

Ferrum Phos.—Inflammation, congestion, heat, fever and pain. Early stage of liver disturbances.

Kali Mur.—Sluggish action of the liver with white or grayish coated tongue and light colored stools. Pains in the region of the liver, constipation, jaundice.

Natrum Sulph.—Symptoms arising from an excess of bile; bitter taste, vomiting of bile or bitter fluid, bilious, greenish stools. Greenish-brown coated tongue, yellow or sallow skin, yellow eyeballs.

Kali Phos.—When bilious attacks are associated with nervousness. Biliousness from worry or excessive mental work. (Use in alternation with other indicated remedy.)

Natrum Mur.—Drowsiness, watery diarrhea, jaundice, associated with catarrh of the stomach.

Dose: Five celloids every one to two hours in acute conditions, at longer intervals after improvement.

SUGGESTIONS

The diet should be light and consist of easily digestible food. The bowels should be kept active with mild laxative remedies or warm water enemas. The drinking of liberal quantities of hot water is also beneficial. If there is fever the patient should remain in bed. If acute condition does not yield to treatment within a reasonable length of time, be sure to consult a reliable physician.

CLINICAL REPORT

Mrs. F. V., 54 years old, five feet three inches in height, 96 pounds weight; parents without clinical interest, good habits, negative Wassermann reaction, urine with biliary pigments and albumin in small quantity. She likes spirituous drinks too well and has the tobacco habit. Her illness before the present time was typhoid fever, frequent colds, and pneumonia.

Present state: Intense fever with sharp pains in the gall bladder region, previously confirmed by expulsion of gall stones. The X-ray examination does not show the presence of the stones.

Symptomatologically she has dizziness, headache,

rise of temperature by access, dirty tongue, fetid breath, dull pain with full sensation in the right hypochondrium, loss of appetite, sour taste of the mouth, icteric coloration in the skin and conjunctivae, and slow and weak pulse.

She suffered from irritability and marked pruritus of the skin. Scanty urine, stools colorless, clay-looking.

I gave Mrs. V. *Natrum Sulph.*, four tablets every two hours, and after one week of treatment I found some relief. Four weeks more of the constant use of the medicine in a smaller dose and the patient recovered.

(Dr. H. L. C.)

BONES, DISEASES OF

CAUSES AND SYMPTOMS

Various mineral salts are, of course, indispensable in the formation, growth and proper nourishment of the bone structure in the human body. A deficiency of *Calcarea Phos.* in the bone cells causes faulty growth and soft bones, and prevents proper repair after fractures.

Calcarea Fluor. and *Silicea* are vital cell-salts in the outer surfaces and coverings of the bones, and a deficiency of these salts is a cause of disturbances in these areas.

The treatment of serious bone diseases such as caries, hip-joint disease, etc., should be entrusted to a competent physician, preferably to one who specializes in this class of disease. In case of a fracture competent treatment is also very important, as improper setting, etc., may cause permanent defects and partial disability.

TREATMENT WITH THE SCHUESSLER REMEDIES

Calcarea Phos.—Weak soft bones in children, tendency to bow-legs, rickets, also delayed, slow dentition. To promote uniting of fractured bones.

Silicea.—Ulceration of the bones; thick, yellow, offensive discharges.

Calcarea Fluor.—Disturbances on the surface of bones, hard, rough elevations. Bruises of the covering of bones.

Ferrum Phos.—Early stage of bone disorders, inflammation, parts are hot, red, swollen and painful.

Dose: In chronic conditions the indicated cell-salt should be given in doses of five celloids three or four times daily. If more than one remedy be indicated, they should be alternated.

SUGGESTIONS

The givıng of *Calcarea Phos.* to a child during the entire period of dentition is considered very beneficial by many physicians.

Hot dressing for injuries to the bones sometimes reduce inflammation and prevent pus formation. After pus has accumulated the services of a physician are usually required to make incision in order to provide for proper drainage of the pus.

CLINICAL REPORT

Miss X., age 16 years. Had fall causing a Colle's fracture. Was treated in a hospital, discharged, pus formed at site of fracture and continued in spite of incision being made and wound drained. After three months patient was brought to me by parent. I prescribed *Silicea* for two weeks. Saw patient again in third week, and the wound had entirely healed.

(R. M. T., M.D.)

BRONCHITIS

CAUSE AND SYMPTOMS

A deficiency of the cell-salts, particularly of *Ferrum Phos.* and *Kali Mur.*, produces in the membranes of the bronchial tubes an excess of albumen and a condition favorable to the lodgment and propagation of infective organisms (bacteria). This produces an infection resulting in an inflammation of the bronchial tubes with fever, distressing dryness of the air passages, hacking cough, difficult breathing with a peculiar crackling, wheezing sound in the chest. As the inflammation progresses the dryness is replaced by secretions of large quantities of mucus, the cough becomes looser and there is a rattling of mucus in the chest. If the lungs or lung covering (pleura) be not involved there is not much pain during the course of bronchitis.

Repeated attacks of acute bronchitis, often following common colds, may result in a chronic bronchitis.

TREATMENT WITH THE SCHUESSLER REMEDIES

Ferrum Phos.—Is the first remedy for the inflammatory conditions, heat, fever and congestion. Short painful cough, without expectoration; short and oppressive breathing. *Ferrum Phos.* should be given in frequent doses in the acute stage, and when additional symptoms appear should be alternated with the remedy indicated by the expectoration, until all inflammatory symptoms disappear.

Kali Mur.—When the expectoration is thick, white, tenacious phlegm, and tongue has a white or grayish-white coating. Alternate with *Ferrum Phos.* when fever is present.

Kali Sulph.—When the expectoration is light-yellow, watery and copious, or greenish, slimy yellow. Alternate with *Ferrum Phos.* when fever is present.

Silicea.—When the expectoration is thick, yellow, and heavy; cough better after warm drinks and aggravated by cold ones. Alternate with *Ferrum Phos.* when fever is present.

Natrum Mur.—Acute bronchitis, with expectoration of clear, watery or frothy mucus. Alternate with *Ferrum Phos.* Chronic bronchitis, "winter cough," with watery symptoms. Phlegm is loose and rattling.

Calcarea Phos.—Expectoration of albuminous mucus (looks like white of egg before it is cooked, not watery). Bronchitis in anemic persons, when the above symptoms are present. Alternate with *Ferrum Phos.* when there is fever.

Natrum Sulph.—When bilious symptoms are present. Alternate with other indicated remedies.

Dose: Two or three celloids for children and five celloids for adults every one-half to one hour, less frequently after improvement and after fever has abated. Use alternately when more than one remedy is indicated.

SUGGESTIONS

The patient must remain in bed as long as there is fever. The diet should be liquid but nourishing. By bringing about free perspiration the fever can usually be reduced—this can be prompted by hot foot baths, hot drinks and the application of hot water bottles.

If there is much mucus the patient should lie with head and shoulders low for a short period at a time, to help gravitation of the secretions.

If dryness of the bronchial membranes is annoying, the inhalation of steam of turpentine in water will usually give relief.

During the entire course of the ailment the patient should wear a soft flannel shirt or cotton jacket.

Bowel regulation is important, if necessary enemas or mild laxative remedies should be used.

CLINICAL REPORTS

Mr. L. had an attack of bronchitis, chills and inflammation with high fever and distressing cough. *Ferrum Phos.* in alternation with *Kali Mur.*, a dose every half hour was taken for 24 hours and then every hour. For sleeplessness a few doses of *Kali Phos.* were given. The improvement was very marked in two days. As the color of sputum was changed to yellow he took *Kali Sulph.* instead of *Kali Mur.* *Natrum Mur.* and *Calcarea Phos.* completed the cure in a little more than ten days.

(Dr. M. M. Z.)

W. C., age 54 years. Telephone man. Works in office and on road, so in and out all the time. A rattling, tight, bronchial cough developed and at times hardly able to talk. No pain and very little white phlegm. Breathing free. Feeling fine every way. Sleep disturbed on account of dry ineffectual cough. *Ferrum Phos.*, three celloids every two hours for a week ended the cough.

(R. C. W., M.D.)

Baby, age 4, with temp. 103, respiration 20, pulse 117. Hard cough, rales all over the chest. *Ferrum Phos.* and *Kali Mur.* alternately. Temperature dropped to 99. Cough better. *Kali Sulph.* was then given and the case was ended.

(L. T. K., M.D.)

BURNS—WOUNDS

The replacement of skin and tissue destroyed by burns and wounds requires a greater supply of mineral salts than the body needs ordinarily. If the burns or wounds are not too extensive the internal use of the Schuessler Remedies will be helpful in maintaining the proper mineral balance.

When a considerable area of the skin and tissue is involved, competent treatment by a physician will do much to prevent or to minimize scarring.

TREATMENT WITH THE SCHUESSLER REMEDIES

Ferrum Phos.—Pain with redness, inflammation of the parts. Also the remedy in cuts, wounds, bruises, etc.

Kali Mur.—Second stage of burns, swelling of parts. A grayish-white exudate appears over the burned surface. After pain has stopped, to assist nature in restoring destroyed tissue.

Calcarea Sulph.—Suppurating burns and wounds, purulent discharge which interferes with healing.

Dose: Five celloids every half hour in the beginning (*Ferrum Phos.*). During the stage of healing every three to four hours.

SUGGESTIONS

The burned or injured surface and parts should be promptly and carefully cleansed with an efficient antiseptic solution. A dressing with gauze bandage after applying a bland healing ointment will serve well for small wounds and burns. When a large area is burned, exclusion of the air from the surface can better be accomplished by spraying or painting hot wax upon the surface.

CLINICAL REPORT

Charles N., aged 10, was sent to help his sister, who was two years his junior, to wash dishes. The two quarreled, and the girl picked up the dish-pan into which she had just emptied a kettle of boiling water, and dashed it upon her brother.

I found the boy suffering from a burn of the second degree, extending almost clear across the breast and from his right shoulder down to a few inches below the hip. The right side of the abdomen was also included. With the greatest caution I opened each blister and allowed the contents to empty. I dressed the parts with vaseline containing 10 per cent carbolic acid. Internally I gave *Ferrum Phos.* alternated with *Kali Mur.* about ten celloids to a half glass of water, of which I gave a teaspoonful dose every hour. There followed a rapid recovery with less scarring than might have been expected.

(DR. A. E. W.)

CANCER

Cancer is a malignant growth of new tissue on the body with a tendency to destroy life. It may occur in any tissues or organ of the body but is more frequent in a few locations. Below are given some of the more common sites for the growth of cancer together with the leading signs and symptoms which suggest the possibility of a cancer being present.

Skin.—Any sore (ulcer) on the skin which persists in spite of treatment to heal it, particularly on the nose or lips, should be suspected as cancer and expert advice obtained. Pipe smokers are particularly likely to develop cancer of the lip or tongue. Any mole or wart that begins to grow or to become sore and irritated after years of quiescence should be watched very carefully and if it does not subside promptly should be removed.

Stomach.—Cancer of the stomach is suspected if stomach symptoms persist with vomiting of material resembling coffee-grounds, burning after eating, vomiting blood and progressive loss of weight and strength. A lump may be felt in the pit of the stomach but rarely is pain felt in the early stages.

Rectum.—Most of the cancers of the intestines are located in the rectum. Progressively increasing constipation with feeling as of plug in the rectum and occasional bleeding or mucous discharge from the rectum are evidences of cancer in this location. These are all early symptoms and should be investigated as only by early diagnosis can cancer be successfully treated.

Breast.—The most common location of cancer in the human body is in the breast. Here the main sign of the presence of cancer is the discovery of a lump in the breast. While benign lumps are found occasionally, at least seven out of ten lumps which are found in the female breast after middle life are cancerous. Therefore a careful investigation of every such lump should be made as soon as discovered in order that treatment may be instituted early.

Uterus.—Any abnormal discharge from the uterus occurring in middle life or after demands that an examination be made to determine the cause. Bleeding from the uterus after the change of life or any watery, foul discharge is very significant of the presence of cancer and should not be neglected but attended to at once, as a few months may allow the cancer to grow beyond the limit of curability.

Early diagnosis and treatment are of greatest importance. Therefore, if the possibility of cancer is suspected, a physician should be consulted at once.

CATARRH

CAUSES AND SYMPTOMS

Catarrh is the term applied to a chronic catarrhal inflammation of the membranes of the nasal passages. The common cold is consequently an acute catarrh; chronic catarrh usually follows repeated acute colds. It causes the membranes to thicken so that they are continually inflamed.

Men engaged in medical research disagree as to the fundamental causes of these most prevalent of all ailments: "Colds and Catarrh." Although the germ or bacteria which some of the scientists believe to produce colds have not yet been isolated by bacteriologists, it seems quite probable that these organisms are real contributory factors in the development of these ills. Some persons, however, seem to be immune to the effects of these bacteria, and others, at certain times only, are unable to resist their invasion and ill-effects.

Whenever there exists in the body a deficiency of the vital cell-salts, it causes a disturbance or abnormal condition somewhere in the system which weakens the resisting power of the body, promoting invasion and propagation of various kinds of bacteria, with resulting ailments such as colds, catarrh, influenza, measles, and many other ills.

(For the treatment of Colds, see page 79.)

TREATMENT WITH THE SCHUESSLER REMEDIES

Kali Mur.—Catarrhal troubles, with white, thick, tenacious phlegm (not transparent). Catarrhs of the head, with stuffy sensations, white or gray coated tongue. Catarrh of any membrane, with a characteristic, white bland discharge.

Natrum Mur.—Catarrhs, with watery, transparent discharges. Catarrhs of anemic people, with frothy discharges, sometimes having a salty taste. Catarrh of any membrane, with above symptoms.

Calcarea Phos.—Is an important remedy in catarrhal affections of anemic persons and chronic cases. Should be given in all cases of catarrh, alternated with other indicated remedies, for its tonic action. Catarrh of any membrane, when the discharge is rich in albumen, transparent, like white of egg before it is cooked. Do not confound the *Calcarea Phos.* discharge with the watery, transparent discharge of *Natrum Mur.*

Kali Sulph.—Third stage of all catarrhs, when the discharges or secretions are yellow, slimy or watery mucus. Thin, yellow discharge from the nose. Generally follows after *Kali Mur.* Symptoms are worse in the evening or in a warm room.

Calcarea Sulph.—Catarrh of any membrane, when the discharge is thick, yellow, mattery and sometimes mixed with blood.

Silicea.—Chronic catarrh with very offensive discharge. Excessive dryness or ulceration of the edges of the nostrils. Itching of the tip of the nose.

Calcarea Fluor.—Stuffy catarrh of the head. Dry coryza. Bronchial catarrh, when tiny, yellow, tough lumps of mucus are coughed up. Diseases of the nasal bones, with very offensive odor.

Natrum Sulph.—Catarrhs, when there is a profuse secretion of greenish mucus. Catarrhs of damp localities, aggravated at every change of the weather.

Dose: The indicated remedy in chronic catarrh should be taken in doses of five celloids, three or four times daily. If more than one remedy is required, take them in alternation.

SUGGESTIONS

The nasal mucous membranes should be kept free from mucus with the use of douches of mild antiseptic or saline solutions. Sprays of bland oily preparations are also helpful in preventing dry mucus adhering to the membranes.

The general condition of the patient must not be overlooked. If in a rundown, anemic condition the patient must be built up by nourishing food and hygienic measures of which fresh air, sunshine and proper exercise are most important. The use of tonic remedies is also beneficial. Constipation, if it is the present, must be corrected. A mild laxative remedy is desirable.

CLINICAL REPORT

A lady over 80 years of age, chronic bronchial catarrh. Extreme exhaustion. Expectoration profuse, yellow. *Kali Sulph.* every two hours. Cured entirely in three weeks.

(W. J. H., M.D.)

CHICKEN POX

SYMPTOMS

This contagious disease, usually confined to children, begins with a general feeling of illness and a slight fever, followed after about 24 hours by a red spotty or pimply eruption, quickly changing into small blisters which soon dry up and crust over. For several days new crops of eruptions appear and go through the same changes.

As a rule chicken pox is not a serious illness, yet the patient should secure proper treatment to prevent serious consequences.

TREATMENT WITH THE SCHUESSLER REMEDIES

Ferrum Phos.—For the inflammatory conditions, heat, fever, pain, restlessness, etc. Alternate with the remedy indicated by the tongue or eruptions.

Kali Mur.—Second stage, after fever has subsided, usually with white or grayish-white coated tongue. Eruptions filled with whitish substance.

Natrum Mur.—When watery symptoms are present; also drowsiness, stupor, etc.

Calcarea Sulph.—When the pustules are discharging thick, heavy, yellow matter.

Kali Sulph.—When the eruption has been checked, alternate *Kali Sulph.* with *Ferrum Phos.* to promote perspiration.

Dose for children: Three celloids of the required cell-salts every one-half to one hour in the early stage, less frequently after the fever has subsided.

SUGGESTIONS

To prevent the spread of the disease the child should be kept from contact with other children. Bathe to keep the skin clean. Protect the child from draughts and cold air which may cause a sudden suppression of the eruption. Prevent scratching to guard against infection and pitting. If necessary, apply carbolized vaseline to allay itching. The diet should be liquid as long as fever is present.

CHLOROSIS

Chlorosis is a form of anemia appearing in young girls at the age of puberty and is recognized by the greenish pallor of the skin. The patient loses weight, is excessively tired, has difficulty in keeping warm, has little appetite, and the appearance of the menstrual period is delayed or irregular.

The treatment with the Schuessler Remedies for chlorosis is given under the chapter of "Anemia." The general treatment is also outlined in the same article.

CHOLERA INFANTUM

(Summer Complaint)

CAUSES AND SYMPTOMS

This acute and serious disease, mostly prevalent during the warm months, attacks chiefly bottle-fed infants. An inability to properly assimilate food, combined with the effects of the heat, brings about a deficiency of some of the vital mineral salts, with the result of a nutritional anemia, aggravated by the ill-effect of even slightly spoiled milk.

Poor general hygienic surroundings and lack of sufficient pure air furnish another background for this disease, which is one of the chief causes of death in infants under one year of age.

Mild cases resemble ordinary diarrhea. In the majority of the cases, however, the disease produces other more serious symptoms. In the beginning the child has vomiting spells, there is some fever, the discharges from the bowels are foul with undigested food at first, then become watery, mixed with mucus. These profuse discharges cause great weakness and emaciation. While the temperature is high the body seems cold, the child is thirsty, but the drinking of water brings on vomiting.

The dangerous character of this illness is the best reason why the treatment should be directed by a doctor. The uses of the Schuessler Remedies with the observation of the highly important general measures as suggested here, while based upon reliable professional experience, are mentioned for emergencies only, when the services of a competent physician cannot be secured.

TREATMENT WITH THE SCHUESSLER REMEDIES

Ferrum Phos.—Is the remedy for the fever; watery, frequent, undigested stools; feverish thirst; vomit of undigested food. Brain symptoms, delirium, rolling of the head, moaning, etc.

Natrum Phos.—When the stools are sour-smelling and green. Cholera infantum, when associated with worms, acid conditions, lack of digestive power, or from eating unripe fruit. Sour vomit, and other acid symptoms.

Calcarea Phos.—One of the most valuable remedies for bowel complaints in teething children, due to non-assimilation of food, or in emaciated children, where the lime salts are deficient. Stools are hot, watery, offensive, profuse and sputtering; sometimes green and undigested.

Kali Phos.—Stools are like rice-water; great depression and exhaustion; stools very offensive and putrid.

Magnesia Phos.—Cholera infantum accompanied with cramp-like pains in the bowels, flatulent colic, drawing up of the legs, convulsions, spurting stools, etc. Alternate with remedies indicated by character of the stools.

Dose: Two or three celloids every 15 to 30 minutes, less frequently after improvements. If more than one remedy is needed, give in alternation.

SUGGESTIONS

During the active stage of the disease, while there is fever and vomiting, the infant should receive no food, especially no milk. When feeding is resumed it should be very dilute and in small quantities at a time until tolerance is well established. White of egg in warm water or dilute barley or oat meal gruel may be given, two teaspoonfuls every two hours. Milk causes fermentation and if given too soon will cause a relapse.

If there is a tendency to irritation from acrid bowel discharges, wash out the colon (enema) with warm water in which a teaspoonful of corn starch has been dissolved. Inject in small quantities but frequently.

Keep the child quiet, the extremities and abdomen protected and warm with flannel, in spite of the high temperature prevailing in this disease.

CLINICAL REPORT

B., 18 months, green watery movements mixed with mucus, with emaciation. Moving its head as if it was too heavy. Pulse rapid, respiration accelerated, complexion of a dirty white appearance.

Ferrum Phos. in hot water every hour for six days and then *Calc. Phos.* in alternation every hour cured the case completely in less than two weeks.

(Dr. M. M. Z.)

CHOLERA MORBUS

CAUSES AND SYMPTOMS

Cholera morbus is an acute disorder of the intestinal tract attended with vomiting, purging of undigested food, abdominal cramps and fever.

A mineral salts deficiency resulting in an improper distribution of water in the system together with the disturbing or irritating effects of spoiled food, unripe fruit or iced drinks upon the gastro-intestinal tract, and a general systemic depression produced by extreme hot weather, are the factors which produce the condition called cholera morbus.

TREATMENT WITH THE SCHUESSLER REMEDIES

Ferrum Phos.—Fever, flushed face; vomiting, watery stools with particles of undigested food; severe cramps.

Natrum Phos.—Excessive acidity, sour smelling stool. Vomiting of sour fluid with curded masses.

Magnesia Phos.—Watery stools expelled with force, with sharp, griping pains in abdomen, relieved by hot applications.

Kali Phos.—Stools have appearance of rice water, very offensive. General depression and nervous exhaustion.

Natrum Mur.—Watery, shiny, frothy stools; transparent glairy slime.

Dose: During the attacks five celloids of the indicated remedy (alternate if more than one remedy is required) every 15 to 30 minutes, at longer intervals after relief is obtained.

SUGGESTIONS

Irrigation of the bowels (enema) is useful in the early stages. At times the patient perspires profusely, and chilling at this stage should be prevented. Cold water should not be taken to satisfy intense thirst; a small quantity of ice quenches the thirst and subdues nausea and vomiting. All food should be stopped during the early acute stage. Food should be taken only when the irritation of the bowels has subsided and then mild, easily digested, semi-solid food should be maintained for a short time.

CLINICAL REPORT

Mr. G., age 45. Because of contamination in the raw milk taken in the evening had an attack of severe diarrhea (cholera morbus) and vomiting the next morning. Exceedingly painful cramps in the calves. Evacuations had the appearance of rice water. *Kali Phos.* celloids every hour, brought relief in six hours. Diet, barley water and orange juice alternately in small quantities. The patient was cured in two days.

(Dr. P. G.)

COLDS

CAUSES AND SYMPTOMS

The medical term for this most prevalent ailment is acute rhinitis (acute catarrh). It is an inflammation of the nasal mucous membranes resulting from a combination of predisposing and exciting causes. The predisposing cause is a depletion of the mineral salts which control the functions of the tissues of the respiratory organs and low general vitality. The exciting causes are a certain form of bacteria which find a breeding place in the depleted tissues; other exciting causes are exposure to cold and wet.

A cold goes through several stages. The first or congestive stage has symptoms of dryness and irritation, sneezing and a burning sensation at times. It is followed by watery discharges from the nose, then by thicker mucus secretions until the disorder gradually disappears, provided that an extension to throat, sinuses, bronchial tubes or lungs does not result.

TREATMENT WITH THE SCHUESSLER REMEDIES

Ferrum Phos.—Early stage, dryness of nose, headache, fever. Alternate with *Kali Mur.*

Kali Mur.—An important remedy in fully developed colds with thick, white secretions, tongue coated white. Alternate with *Ferrum Phos.*

Natrum Mur.—Watery, clear, frothy discharge, sneezing, irritated nostrils, cold blisters on lips, loss of sense of smell.

Kali Sulph.—Dry skin, fever, to promote perspiration (use also *Ferrum Phos.*); profuse discharge of greenish-yellow mucus.

Dose: Five celloids of the indicated remedy every one-half to one hour, less frequently after condition has improved.

SUGGESTIONS

A cold, no matter how slight, should never be neglected. If there is congestion and fever the patient should remain indoors. In fact, remaining in bed for one or two days is considered to be almost an infallible prophylactic measure against complications. The use of nasal douches and gargling with antiseptic solutions is helpful, but probably of greater use as a preventive measure than as a curative treatment. The regulation of the bowels, with mild laxative remedies, if necessary, is desirable. Patients, especially children, who are subject to habitual frequent colds, should be encouraged to spend as much time as possible in the open air, bathe in the morning in cold water, and to sleep with plenty of ventilation in the room. The nose should be examined for obstructions, particularly adenoids, if formed, such measures as necessary to insure a free nasal tract should be instituted. The use of *Calcarea Phos.* and *Ferrum Phos.* in alternation by weakly or anemic patients is useful when there is a predisposition to take frequent colds. Soreness and irritation can be relieved by the application of Menthol Balm to the nostrils and nasal membranes.

COLIC

CAUSES AND SYMPTOMS

Colic is a sharp abdominal pain arising from various causes. It is, as a rule, a symptom associated with disturbances such as diarrhea, cholera infantum, intestinal disorders, menstrual troubles, gall-stones, etc.

The most satisfactory treatment obviously is one directed towards the removal of the cause, which in some cases can be readily discovered.

TREATMENT WITH THE SCHUESSLER REMEDIES

Magnesia Phos.—The remedy for the pain, in alternation with the remedy indicated by the symptoms which gave rise to the pain. Colic of infants, with drawing up of the legs. Pain relieved by bending double. Flatulent colic, eased by friction, heat or belching of gas. Colic coming and going by spells. Pains are crampy and constrictive, eased by heat.

Calcarea Phos.—If *Magnesia Phos.*, though indicated, fails to give relief, follow with *Calcarea Phos.* Colic due to non-assimilation of food, or in teething children when the lime salts are deficient.

Natrum Phos.—Colic of children, with worms or symptoms of acidity, green, sour-smelling stools, vomiting of curdled milk, etc.

Natrum Sulph.—Bilious colic, with vomiting of bile; bitter taste in the mouth and brownish-green coating on root of tongue. Lead colic: should be given frequently.

Ferrum Phos.—Menstrual colic, with fever, quickened pulse, etc., in alternation with *Magnesia Phos.*

Kali Sulph.—Is frequently useful after *Magnesia Phos.* if the abdomen feels cold, tense, distended from gas, or if the colic is due to excitement and sudden chill shortly after.

Dose: Magnesia Phos. should be given every 15 to 30 minutes, less frequently after relief is obtained. The other indicated remedies every one-half to one hour in alternation with *Magnesia Phos.* The dose for adults is five celloids, for children two to three celloids.

SUGGESTIONS

The proper general treatment depends upon the disorder with which the colic is associated, and the proper suggestions are given under the respective chapters. Generally speaking, rest in bed and hot applications upon the abdomen will prove beneficial and procure relief. If the bowels are not active a warm water enema with a teaspoonful of salt is recommended. Do not give laxative remedies until sure that no inflammation is present. Observation of a careful, restricted diet (liquid or semi-liquid) is necessary until all symptoms of disturbance have ceased.

CLINICAL REPORT

Was consulted in regard to William M., 8 years old, for colic that came on at night. Gave him *Magnesia Phos.* with instructions that he should take two tablets dissolved in hot water every ten minutes in case he had colic. His mother told me that the third dose was not needed, but she gave it anyway.

(Dr. Thos. A. B.)

CONSTIPATION

CAUSES AND SYMPTOMS

Constipation is the condition of delayed, difficult or deficient movements of the bowels. There are numerous contributing causes to this prevalent and health-disturbing condition. In many cases we find an irregular distribution of water in the digestive and intestinal tract. Disturbance of the nervous mechanism which controls intestinal activity is another cause. Another prominent factor is the modern tendency to highly refined foods, which lack much of the undigestible bulk, necessary to provide stimulation to the intestines. Such foods are also deficient in the natural mineral salts which the body requires.

Bad habits as to regularity of eating and drinking, and inattention to the calls of nature play an important part in the development of chronic constipation.

TREATMENT WITH THE SCHUESSLER REMEDIES

Ferrum Phos.—Constipation accompanied by a feeling of heat in the rectum, causing absorption of the natural fluids of the feces, and resulting in hardening and drying of the discharges. Piles, prolapsus of the rectum, inflammations and fever are frequently associated with this type of constipation.

Kali Mur.—Constipation, with light-colored stools, from torpidity of the liver and want of bile. With white or grayish-white coated tongue, or when fat foods disagree.

Kali Phos.—Stools dark-brown, streaked with yellowish-green mucus. Constipation due to sedentary habits or excessive nervous conditions, or mental strains.

Natrum Mur.—Constipation when caused from lack of moisture in the intestines. Dryness of the bowels, with watery secretions in other parts, watery eyes, excess of saliva, watery vomiting, etc. Constipation with water-brash; dull, heavy headache; hard, dry, black lumpy stools, difficult to pass; torn, smarting feeling after stool.

Natrum Sulph.—With bilious symptoms, hard, knotty stools; or soft stools, difficult to expel.

Calcarea Fluor.—Inability of muscles to expel feces requires this remedy. The muscles of the rectum become relaxed, allowing a too large accumulation of fecal matter.

Calcarea Phos.—Hard stool, with occasional pieces of albuminous mucus, also in anemic patients. Constipation in the aged.

Natrum Phos.—Constipation of infants, with occasional attacks of diarrhea. Sour or acid conditions.

Dose: Five celloids of the indicated remedies three or four times daily. Use in alternation if more than one remedy is required. In chronic conditions the remedies must be taken for a considerable length of time.

SUGGESTIONS

The use of the Schuessler Remedies, valuable as they may prove in correcting some of the underlying causes of constipation, are only a part of the treatment required to correct chronic constipation.

The diet must include substantial quantities of food which leave sufficient residue (roughage) to stimulate the bowels to muscular activity. Fruits and vegetables are particularly desirable for this purpose. Exercises involving the muscles in the abdomen are also desirable.

The patient should establish regularity in time of bowel movements. About one-half hour after breakfast a daily attempt should be made to move the bowels and persisted in even if unsuccessful at the beginning. This is an important factor in the correction of constipation.

The daily use of laxatives or purgatives is undesirable because it does not cure constipation and discourages bowel movements from natural impulses. Persons who have become accustomed to the use of laxative remedies cannot, however, abruptly discontinue this practice without danger of causing severe auto-intoxication.

By following the suggestions given in this article and with the persistent use of the indicated Schuessler Remedies the dosage and frequency in taking laxative remedies can be cut down gradually and finally be discontinued entirely.

The occasional use of a dependable, mild laxative remedy, in acute cases of constipation, as may occur from errors in diet, or in the beginning of minor ailments, is advisable and will bring desirable relief.

CONSUMPTION

(Pulmonary Tuberculosis)

Consumption or tuberculosis is caused by infection by the "Bacillus Tuberculosis." This organism reaches the lungs by inhalation where it will multiply if conditions are favorable, form inflammatory nodules gradually destroying the tissues of the lungs. The "Bacillus" is present in the sputum and discharges of tubercular patients and is capable of living for a long time in a dried state. A predisposing cause of tuberculosis is the demineralization of the body, especially of the Calcium (Calcarea) salts. It permits accumulation of waste material to be thrown off by the lungs, and this material forms a favorable medium for the growths of the germs of tuberculosis.

Treatment of this serious disease should always be under the supervision of a physician. The hope of a complete cure depends very much upon an early diagnosis and treatment.

The early symptoms are weakness, loss of weight, slight cough and daily rise of temperature (fever) which may be only slight, requiring a fever thermometer to be detected. The patient appears anemic, and as the disease progresses becomes weaker, the fever becomes more active, there is profuse sweat, especially at night.

CONVULSIONS—CRAMPS—SPASMS

CAUSES AND SYMPTOMS

Certain cell salts, especially *Kali Phos., Magnesia Phos.* and *Calcarea Phos.,* are the vital elements which maintain the nerve tissues and control their functions.

A deficiency in these salts is the basic cause of irritability of the nervous system. So-called reflex excitation of certain nerves results in muscular spasm in the area of that nerve and in some cases affecting the entire body. Such conditions are common in children whereas in adults convulsions generally have a more serious organic disease as the exciting cause. They may be symptoms of epilepsy, chorea, etc.

Generally, convulsions or cramps occur suddenly. They are easily recognizable by the involuntary contractions of the muscles, either with more or less prolonged rigidity of the muscles or with rapidly repeating jerking of the muscles.

Patients who are periodically or repeatedly afflicted with attacks of convulsions should submit to thorough examination and treatment by a competent physician.

TREATMENT WITH THE SCHUESSLER REMEDIES

Magnesia Phos.—This is the chief remedy for spasms in any part of the body. Convulsions, twitchings, contractions, cramps, fits, writers' cramp, twitching of facial muscles, spasmodic stammering, squinting, jerking of limbs.

Calcarea Phos.—After or in alternation with *Magnesia Phos.* for spasmodic affections in teething children or in cases where the lime salts are deficient, as is generally the case in anemic subjects.

Kali Phos.—Fits and spasm from fright, excitement, with pale or livid countenance. Hysterical spasms.

Ferrum Phos.—For the febrile conditions which frequently accompany spasms, especially in teething children.

Dose: Two to five celloids of the indicated remedy every 15 to 30 minutes during attacks, less frequently after relief is obtained. During period of dentition give *Calcarea Phos.* three or four times daily.

SUGGESTIONS

Spasms of children frequently originate from stomach and intestinal disturbances, and a warm water enema will help to clear the intestines. To bring about relaxation in case of convulsions, a warm bath will be helpful, wrap the child in warm blanket after bathing. When the child is able to swallow, give a teaspoonful of castor oil to rid the tract of food poisons.

CLINICAL REPORT

A young couple called at my office with a little five months' old baby. They stated that she would lie on the bed and all of a sudden would cross her legs and fold her arms across her chest, and then would go into what appeared to be a spasm, then would break out into a profuse sweat. I prescribed *Calcarea Phos.* and asked the parents to call again in thirty days.

I lost track of the case for nearly a year when a brother of this family called with a four-year-old girl. He stated this girl was like his brother's, except that instead of lying down she would sit on the table with her legs hanging down and would swing them back and forth and likewise break out into a sweat.

About five months later another couple brought a two-year-old child with the same profuse sweat and similar symptoms. I prescribed *Calcarea Phos.* in these cases with very good results.

(DR. A. B. H.)

COUGH

CAUSES AND SYMPTOMS

Cough is a symptom associated with various disorders of the respiratory tract. It may result from an irritation of the membranes of the tract or arise from an effort to rid the passages from an excess of mucus. The relation of mineral deficiencies to the various disorders of the respiratory ills, such as bronchitis, pneumonia, colds, whooping cough, croup, etc., is given under their respective titles in this book.

TREATMENT WITH THE SCHUESSLER REMEDIES

Kali Mur.—Loud, noisy and spasmodic coughs, accompanied with white or grayish-white coated tongue; the expectoration is thick, milky-white, tenacious. Croupy, hard cough; croup-like hoarseness.

Ferrum Phos.—Short, acute, painful cough, with soreness in the lungs and no expectoration. Tickling cough, caused by irritation of the bronchial tubes. Hard and dry cough, with soreness of chest. For the inflammatory symptoms accompanying a cough.

Magnesia Phos.—Paroxysms of coughing, without expectoration. Spasmodic cough, loud and noisy, like whooping cough; relieved by hot drinks.

Kali Sulph.—Late stage of inflammatory coughs, with expectoration of slimy, yellow or watery-yellow matter. Always worse in a warm room or in the evening; better in cool, open air. Hard, hoarse, croupy cough, with rattling of mucus in the chest.

Calcarea Sulph.—Cough, when the expectoration is loose, mattery and sometimes streaked with blood.

Silicea.—Chronic coughs, with thick, profuse, yellowish-green, mattery expectoration; always worse in the morning on rising or on lying down at night; worse from cold drinks.

Natrum Mur.—Cough, with clear, watery expectoration, sometimes tasting salty, or with excessive discharge of watery secretions from the eyes, nose or mouth. In dry, tickling, hacking, irritating coughs.

Calcarea Phos.—Expectoration clear but thick, rich in albumen—like the white of an egg before it is cooked. Intercurrently in all chronic coughs.

Dose: In acute coughs the indicated remedies should be taken in doses of five celloids every hour, less frequently after relief is obtained. If more than one remedy is indicated, their use in alternation is recommended.

SUGGESTIONS

Acute coughs usually cease as soon as the irritation of the membranes is allayed. Smoking and alcoholic beverages have a tendency to aggravate the irritation. If the cough cannot be controlled within a reasonable length of time a thorough medical examination is recommended in order to correctly diagnose the underlying disorder.

CLINICAL REPORTS

Child, 10 years. Had obstinate cough of three months' duration. Mother brought patient to me a week ago. Said home remedies did no good. Radio advertised remedies were a failure. Put patient on *Kali Mur.*, two tablets every two hours. Cured this case in one week.

When there is soreness, a raw feeling in the chest and a tickling in the throat with phlegm, hawking of mucus from the posterior nares, *Kali Mur.* is the remedy.

Give three tablets every two hours. It is an ideal cough remedy with the raw feeling in the chest and tickling in the throat.

(Dr. A. S.)

One morning a gentleman stopped in my office to see if I could help his cough which had bothered him every spring and fall for twenty years, always accompanied with a slight fever. He related to me how he had gone to his family physician three times and each time got quinine tablets which did no good.

After questioning him for a few minutes I gave him a few celloids of *Natrum Mur.* and prepared him a vial of this remedy with orders to take every four hours.

When the medicine was consumed he called again to ask what was in the medicine. He said, "Why, it worked like magic. I had never given Homeopathic or Schuessler medicine any thought. Considered them just remedies for children, but you have been treating my wife and daughter so successfully I concluded to give these remedies a trial and must say am wonderfully pleased."

(E. B. F., M.D.)

CROUP

SYMPTOMS

Croup is a spasmodic affliction of the larynx, and is a common disease of childhood. The early signs of croup are similar to those of a head cold, but the cough is dry and has a barking or ringing sound. Spasmodic attacks of difficult breathing follow, frequently at night, they subside in a few hours, but have a tendency to recurrence. Alarming as it may appear, spasmodic croup is seldom serious, the spasm passing off after a time. The exudation of fibrine from the membranes during the course of the ailment point to a *Kali Mur.* deficiency.

Formerly a distinction was made between spasmodic croup, described above, and true membraneous croup, but the latter is now properly designated as laryngeal diphtheria (see diphtheria), a dangerous disease not suitable for home treatment.

TREATMENT WITH THE SCHUESSLER REMEDIES

Kali Mur.—Is the chief remedy in croup, to control the exudation of fibrine, and acute, spasmodic cough. Should be given in alternation with *Ferrum Phos.* for the febrile symptoms (fever).

Ferrum Phos.—For the fever, early stage, dry hard cough, shortness of breath, etc. Alternate with *Kali Mur.*

Kali Phos.—When treatment is delayed and there is danger of a collapse; countenance is pale or livid; nervous prostration. Give in alternation with *Kali Mur.*

Magnesia Phos.—Suffocative cough, gasping for breath, spasmodic closure of the throat, sudden shrill voice.

Dose for children: Three celloids every 15 minutes or one-half hour, less frequently after breathing has become easy.

SUGGESTIONS

Application of a cold compress about the throat, changed as soon as it becomes warm, is very helpful to relieve attacks of croup. If the water is not very cold a piece of ice may be put in the pan. Cover the cold compress with a dry flannel wrapped around the throat. The room should be warm and the air moistened. A kettle of boiling water may be placed near the bed for this purpose, and if the attack is severe a sheet may be draped over the child's head to make a steam tent, and the steam from the kettle directed under it. Protect the child from drafts and chills.

Following the attack the child should be kept quiet, as exercise or excitement is likely to bring a recurrence of the croup the following night.

CLINICAL REPORT

Loren M., age 3. Was called about 2 a. m. Found him almost choking and gave *Kali Mur.* celloids in combination with *Ferrum Phos.* celloids for the febrile symptoms. He got up in the morning and played around, and the croupy cough was entirely gone by nightfall.

(Dr. A. C. N.)

CYSTITIS

(Inflammation of the Bladder)

CAUSES AND SYMPTOMS

A disturbance of the mineral balance in the system disorganizes the normal activities of the cells which make up the structure of the human body. It also weakens or destroys our natural powers of resistance to disease. In a normal state of health the system offers effective resistance to various disturbing influences, such as the ever present harmful bacteria, exposure to the elements, faulty diet, etc.

Bladder inflammation has various exciting causes, among which infections by organisms is most common.

The principal factor, however, is the weakened condition of the bladder membranes arising from mineral salt deficiencies.

The principal symptoms of cystitis are pain, inflammation, congestion, difficulty in passing urine, which is cloudy. The severity of the symptoms in some cases necessitates the use of the catheter and other treatment which demand the services of a competent physician.

TREATMENT WITH THE SCHUESSLER REMEDIES

Ferrum Phos.—First stage with fever, frequent urination and burning pain. Urination difficult, suppressed with constant urging.

Kali Mur.—Second stage with swelling of the tissues and thick, white mucus in the urine. Also in chronic cystitis with dark red urine.

Kali Phos.—Cystitis when associated with nervousness, prostration. Scalding urination, cutting pain.

Magnesia Phos.—Ineffectual and painful straining, urine passes in drops. Severe spasmodic pains.

Calcarea Sulph.—Chronic cystitis with pus in the urine.

Dose: During acute attacks five celloids of the indicated remedy every one-half to one hour, less frequently after relief is obtained. In chronic cases five celloids three or four times daily.

SUGGESTIONS

The diet should be bland non-irritating. Highly seasoned food, tea, coffee and alcoholic beverages must be eliminated. Milk should be the principal food, especially during the acute condition.

Pain and difficult urination can sometimes be relieved with applications over the region of the bladder, of cloths wrung out in hot water.

In severe cases, with much pus in the urine, irrigation of the bladder and in some cases continuous drainage may be required. Application of these measures require the services of a physician.

CLINICAL REPORTS

Man, age 44. Mr. S. came to me stating that he had kidney trouble for six years. On examination found that pain over bladder in left side by urethra on pressure. Laboratory test showed normal except fungus in urine.

Placed this man on *Kali Mur.* and *Calcarea Fluor.*, five celloids of each before meals, and in one month was rid of pain and laboratory test did not show any more fungus. That was three years ago and the man is all right today. He still takes a dose of the two above remedies about once per week, he tells me.

(Dr. F. H.)

Mr. S., age 45 years. Had gonorrhea long ago which was followed by stricture with great "forcing." The urine which passed with difficulty per urethra showed pus and mucus, bladder tender to pressure. I prescribed *Silicea* for ten days, then *Calcarea Sulph.* In four weeks he passed his urine normally. I gave *Kali Mur.* for the cystitis for ten days which entirely cleared the urine of mucus, etc.

(R. M. T., M.D.)

DEAFNESS

(See Ear Diseases)

DEBILITY

CAUSES

Debility is a condition of general weakness and lack of strength. Frequently it is due to anemia or to a deficiency of cell-salts as a result of disturbance of digestive functions or of some constitutional disease. A treatment directed towards the elimination of the causative factors, as a rule, produces the most desirable results.

TREATMENT WITH THE SCHUESSLER REMEDIES

Calcarea Phos.—An important remedy in debility generally, anemic or run-down condition, loss of weight.

Kali Phos.—In cases where the nervous system is involved; mental depression. Alternate with *Calcarea Phos.*

Dose: Five celloids every three hours.

SUGGESTIONS

Obviously a proper, well-balanced nutritious diet is of considerable value and so is the use of a good Tissue Tonic. Out-door exercise and exposure to sunshine also have a beneficial influence. For further suggestions in general treatment, see articles under titles of diseases which may be associated with debility, such as anemia, indigestion, consumption, neurasthenia, etc.

DELIRIUM

CAUSES AND SYMPTOMS

Delirium is a temporary affection of the brain resulting in delusions, hallucinations and disconnected speech.

It may be the result of a congestion or a fever, or the exciting cause may be a poison introduced into the body or generated within the body, as in intestinal auto-intoxication. Delirium tremens may follow excessive drinking of alcohol. The underlying cause usually is a disturbance of the distribution of water in the system, consequent upon a deficiency of the cell-salt *Natrum Mur.*

In the course of fevers, especially in children, the congestion and pressure upon the brain clouds the mind and produces delirium.

The possibility of serious complication demands that the patient should receive treatment under the personal supervision of a physician.

EMERGENCY TREATMENT WITH THE SCHUESSLER REMEDIES

Natrum Mur.—Delirium with muttering and wandering; stupor and sleepiness. Delirium with frothy bubbles of saliva on the tongue.

Ferrum Phos.—During acute fevers, flushed face, rapid pulse, rush of blood to the head.

Kali Phos.—Fear, hallucinations in delirium tremens, rambling talk.

Dose: Five celloids at frequent intervals, every 15 minutes if possible. Continue the use of the remedy for several days after the delirium has ceased, giving five celloids three or four times daily.

SUGGESTIONS

In delirium attending fevers a cool bath, 68° F., may be given and followed by an ice bag or cold compress applied to the head. The patient should be kept quiet.

DENTITION

(Teething)

One of the vital requirements for the formation of good, healthy teeth is the presence of an adequate supply of the cell salt phosphate of lime (*Calcarea Phos.*). A deficiency thereof will result in delayed dentition and other disturbances such as faulty structural development. Such conditions are most frequently found in artificially fed infants because of the lack of a sufficient amount of this salt in the food. Another important salt entering into tooth formation is *Calcarea Fluor.*, which is the main ingredient of the hard enamel surface of the teeth. Teeth are easily broken and decayed when the enamel is not intact.

APPLICATION OF THE SCHUESSLER REMEDIES

Calcarea Phos.—Teething late or retarded. Bones in child are soft, fontanelles open. Disturbance of digestion with diarrhea. If the mother during the latter months of pregnancy should take *Calcarea Phos.* three times daily, delayed or faulty dentition could, to an appreciable extent, be prevented. In delayed dentition two or three celloids of *Calcarea Phos.* should be dissolved in each bottle of milk given to the baby.

Ferrum Phos.—Gums swollen and hot with fever. Face flushed, infant is restless and irritable (two celloids every hour).

Magnesia Phos.—Convulsions associated with teething; spasmodic colic with diarrhea (three celloids in hot water every 15 minutes).

Calcarea Fluor.—Teeth rough and crumble easily. Deficient enamel; vomiting during dentition (three celloids four times daily).

Natrum Mur.—Constant excessive drooling of saliva during dentition, even while asleep (three celloids four times a day).

Silicea.—Slow, painful dentition, much perspiration about the head.

SUGGESTIONS

The importance of the temporary teeth is often overlooked as they are a guide to the type of permanent teeth that may be expected, as well as furnishing support to the growing jaws and defining their shape.

The time of the temporary teeth is about as follows:

Middle lower incisors......	6th to 8th month
Upper central incisors.....	8th to 12th month
Upper lateral incisors.....	10th to 12th month
Lower lateral incisors......	12th to 15th month
Four anterior molars......	14th to 16th month
Four canines..............	18th to 20th month
Four posterior molars.....	20th to 30th month

If no teeth are erupted by the end of the first year it is usually an evidence of rickets, which should receive proper treatment.

Bottle-fed babies are usually slower in cutting teeth. Caries (decay) of temporary teeth affect the oncoming teeth and should be cared for promptly.

CLINICAL REPORTS

Willie A., aged 7 years. This child had the appearance of an undernourished child of four years, had lost all his teeth with no replacement.

I put him on *Silicea* and *Calcarea Phos.*, three celloids every three hours. As they were going back to their home in the East I gave them enough celloids to last a month and instructions to report results at the end of that time and I would send more celloids if needed.

I did not hear from them in six months and then they came back to stay, and the case was once more placed in my hands. The boy showed a marked improvement in health but no sign of any action in the gums.

I again placed him under the same treatment, and in two months the boy cut his first tooth and was rugged and healthy.

At this time I quit active practice and moved to Florida, so cannot report final results.

(Dr. Geo. O.)

I have had many patients who have trouble cutting wisdom teeth, causing swollen gums, sore throat, and sore gland on neck, hard to swallow. I have them use *Silicea* which stops these conditions and lets the teeth come through without lancing, etc., a thing I never have had to do.

(L. O. H., D.D.S.)

I was called to see a baby seven months old who had been under the care of another physician for two months. I found it fretful, feverish, gums terribly swollen; didn't have a tooth. I gave the mother an ounce of *Calcarea Phos.*, instructing her to give two celloids every two hours. I also left with her *Ferrum Phos.*, to be given in alternation for 24 to 36 hours. The mother called, as per instructions, in 10 days, and reported the baby had four teeth and had improved in every way. I gave her another supply of *Calcarea Phos.* and told her to keep up the celloids which she did. She now sends every mother with a teething baby to me. (DR. F. G.)

DIABETES

Diabetes is a constitutional disease due to degeneration of certain portions of the pancreas resulting in the inability of the body to assimilate sugar, as a consequence of which the sugar content of the blood is raised. This excess of sugar in the blood is largely thrown off by the kidneys, thus presenting one of the characteristic findings in this disease, sugar in the urine.

Diabetes is a very dangerous disease and one the care and management of which is not suitable for home treatment, but should always be under the supervision of a physician. The cardinal symptoms are outlined below for the purpose of making recognition easy, that immediate examination may be made and proper treatment instituted early:

Increased thirst with an increase in the daily output of urine (three quarts or more), increased appetite with a gradual loss of weight and strength, dryness of the skin and itching. When any or all of these symptoms appear examination should be made for diabetes.

DIARRHEA

CAUSES

The predisposing cause of diarrhea is found in a mineral salt deficiency. The fluid control of the body rests largely in the sodium salts and a deficiency thereof may result in the excretion of water from the bowel. Other salt deficiencies may give rise to digestive disturbances resulting in diarrhea. There are other exciting causes of diarrhea, such as unripe fruits, decayed foods, iced drinks, etc. Bacteria or parasites may produce an infection which causes severe and persistent diarrhea. (See also Dysentery and Typhoid Fever.)

TREATMENT WITH THE SCHUESSLER REMEDIES

Ferrum Phos.—Diarrhea of sudden onset with fever, thirst. Diarrhea caused by a chill; stools consist of undigested food, or watery, frequent stools, pain.

Kali Mur.—Stools are light-colored, pale yellow. White or slimy stools after eating rich food, with white-coated tongue. Bloody or slimy stools.

Kali Phos.—Diarrhea with foul-smelling, putrid stools; discharges like rice-water; offensive stools, with or without pain; also when depression or exhaustion is present.

Kali Sulph.—Stools are yellow, watery, and mattery; tongue coated light yellow; sometimes cramps in the bowels.

Natrum Mur.—Stools are watery, slimy, transparent or of glairy slime, caused by an excessive use of salt; stools cause soreness and smarting.

Natrum Sulph.—Stools are mattery, dark or green, bilious. Chronic diarrhea, with loose watery morning stools; worse in cold, wet weather. Diarrheas of old people.

Natrum Phos.—Stools are sour-smelling and green, due to acidity of the stomach and bowels. Summer diarheas from eating unripe fruit, or associated with worms. Diarrheas of teething children, with acid symptoms, and creamy, golden-yellow coating on the tongue.

Calcarea Phos.—One of the best remedies for diarrheas of teething children, due to poor assimilation of food; should be alternated with the remedy indicated by color of the stool. Stools are hot, watery, offensive, profuse and sputtering; sometimes green or undigested. Diarrheas in pale, anemic or rachitic children.

Magnesia Phos.—For cramp-like pain in the bowels, flatulent colic; relieved by hot applications. Alternate with the remedies indicated by color of the stool.

Dose: During the early, acute stage five celloids every hour, then every two to three hours. If more than one remedy is required they should be taken in alternation.

SUGGESTIONS

As a general rule no food should be taken during acute attacks of diarrhea. If vomiting and thirst is excessive, enemas of warm water should be given every three or four hours. This will wash out and relieve the irritation of the bowel.

When feeding is resumed, begin with very light, easily digested foods and at the first sign of recurrence of the diarrhea stop feeding again. As most diarrheas are aggravated by chilling, the patient should be kept warm and quiet.

The discharges should not be checked too rapidly by the use of narcotics or other drastic drugs. The restoration of the equilibrium of the cell-salts is the safe and natural way to correct the trouble.

If the diarrhea persists a careful medical examination is recommended in order to discover the possible presence of a more serious underlying trouble.

DIPHTHERIA

Diphtheria is a very dangerous and highly contagious disease especially affecting children. It is not a disease suitable for home treatment and should be under a physician's care as soon as possible, as the earlier treatment is started the better the hope of recovery. Any case showing signs of diphtheria calls for immediate examination of the throat and if deemed advisable a swab should be taken from the throat for bacteriological test. If your physician is not available for this examination, the health authorities of your community or township will make the necessary tests.

Immediately upon diagnosis of diphtheria being made quarantine regulations must be established and the patient carefully isolated from other members of the family except the nurse, who must be very careful to wash her hands and gargle her throat—Creozone Antiseptic (Luyties) is excellent for this purpose—as often as she comes into close contact with the patient. All discharges from the throat, urine and stools as well as all articles coming into contact with the patient must be destroyed or carefully sterilized by boiling. For purposes of recognition the following description of the symptoms of diphtheria is given.

After exposure the first symptoms appear in from two to five days. The disease usually starts with a chill or vomiting, headache and sore throat or difficulty in swallowing. The fever is not high, 101-102.° The throat is reddened and somewhat swollen, usually more on one side than the other, and within 12-24 hours a grayish or yellowish membrane appears on the tonsil or the soft palate. This membrane rapidly spreads and toughens and becomes firmly fixed to the underlying mucous membrane so that it cannot be easily rubbed off, and if it is pulled off leaves a raw bleeding surface. The membrane rapidly reforms when torn off. With the development of the disease the membrane may extend over the palate and into the nose or downward into the larynx causing membranous croup. A barking cough and suffocating are evidence of involvement of the larynx. There is danger of the child choking to death on this membrane. Toxic symptoms appear on the second or third day when the disease is severe and there is danger of collapse. The patient must be kept flat in bed and not allowed to sit up till convalescence is well established. Failure to observe this precaution has been the cause of sudden death from collapse and heart failure.

In order to preserve the strength of the patient a light and easily digested diet must be given at frequent intervals if the patient is able to swallow. For this purpose broths, milk alone or with raw egg beaten into it, and custards are acceptable. Give an abundance of water.

DIZZINESS

(See Vertigo)

DROPSY

Dropsy is an accumulation of water in any cavity or in the tissues of the body.

The relationship between the sodium salts and the water in the body is now clearly established. *Natrum Phos.* controls the water intake; *Natrum Mur.*, the distribution in the body; and *Natrum Sulph.*, the elimination of water from the body. Disturbances of the water element in the human body are frequently due to failure of one of the sodium salts to perform its duty in water control.

In any condition of mineral salt deficiency tending to produce dropsy, any stagnation of the blood stream will assist in the escape of water from the vessels into the tissues. Such a condition may arise in heart disease with weakened heart impulse, in the condition of hardening of the arteries, where loss of elasticity in the vessel walls prevents the forward motion of the blood stream, and also in chronic kidney diseases due to failure of the kidneys to eliminate sufficient water and causing a waterlogging of the body.

Due to the fact that dropsy is frequently a symptom of serious disorders, such as mentioned in the preceding paragraph, the proper treatment demands a careful diagnosis, and should be conducted under the supervision of a competent physician.

DYSENTERY

CAUSES AND SYMPTOMS

Dysentery is an inflammatory disease of the intestines dependent on infection with organisms or bacilli, through the medium of drinking water or food infected with this organism. Predisposing to the infection is a deficiency in certain of the mineral salts which renders the bowel susceptible to the eruption and growth of the organisms. The weakening of the tissues of the bowel wall due to an iron deficiency allows hemorrhage to occur, while the lack of *Kali Mur.* is responsible for the excess of albumin in the bowel which is the medium upon which the infecting organism thrives.

The onset of dysentery is sudden and severe. Pain is intense and there is much ineffectual straining at stool. The stools consist almost entirely of mucus and blood and rapidly increase in frequency. Fever is fairly high with intense thirst and prostration is common. It is a dangerous disease and relapses or recurrences are common. Treatment should always be conducted under supervision of a physician and the suggestions given here are only intended for use in emergencies.

EMERGENCY TREATMENT

Magnesia Phos.—For the pains and cramps in the bowels and abdomen. Pains in the rectum, with constant urging to go to stool. Pains better from bending double; spasms.

Ferrum Phos.—In the commencement of the disease, and for the fever, inflammatory pain, etc. Stools hot and watery. Alternate with *Kali Mur.*

Kali Mur.—If alternated with *Ferrum Phos.,* in the early period of the disease, it will frequently serve to relieve the symptoms, if taken in time. Intense, cutting or steady pain in the bowels; constant urging to stool, with pain and purging. Pale-yellow slimy stools, white-coated tongue, etc. Take also *Magnesia Phos.,* for the pain.

Kali Phos.—Stools offensive and putrid. At times the stools are pure blood, tongue dry, abdomen swells, delirium sets in and the discharges are putrid.

Dose: Five celloids of the indicated remedy every two hours. Usually alternate *Ferrum Phos.* and *Kali Mur.* with *Magnesia Phos.* taken in frequent doses (every one-half hour in hot water) when there is pain.

SUGGESTIONS

The diet should be liquid, milk being preferable in most cases. When water is given it should be boiled. To offset the great loss of liquids an abundance of liquid should be given to the patient. The bowel may be washed out once or twice daily with warm water in which one teaspoonful of corn starch has been dissolved. Hot water bottles may be applied to ease the pain. After recovering return to a regular diet gradually.

CLINICAL REPORT

A mechanic of robust frame was taken with acute bacillary dysentery. The usual remedies gave no relief. Though suffering, it was impossible to keep him at rest in bed. Finally, *Ferrum Phos.,* five celloids four times a day, gave notable relief, especially to passage of blood. He recovered with no other remedy except very small doses of ipecac.

(M. L. K., M.D.)

DYSMENORRHEA

(Painful Menstruation)

CAUSES AND SYMPTOMS

Dysmenorrhea is the medical term applied to the painful symptoms which may be present during the menstrual period. Very frequently it is caused by a congestion and inflammation of the pelvic organs; it may be neuralgic, from irritability of the nervous system. Another cause is a possible obstruction to the menstrual flow. This type is relieved as soon as the flow is well established whereas the congestive type may increase during the flow.

The pains are sometimes accompanied by nausea, vomiting, headache, backache.

TREATMENT WITH THE SCHUESSLER REMEDIES

Magnesia Phos.—For the spasms of pain. Cramps, labor-like, bearing-down pains. Sharp cutting pain relieved by heat.

Ferrum Phos.—Painful menstruation, with bright red flow, flushed face and quickened pulse. Vomiting of undigested food. Congestion of the pelvic organs begins several days before the flow. Alternate with *Magnesia Phos.* during the attack.

Kali Phos.—Menstrual colic in pale, tearful, irritable, sensitive women, weakness of the nervous system; flow deep, dark red. Alternate with *Ferrum Phos.*

Calcarea Phos.—Intercurrently in anemic patients In girls at puberty, scanty flow.

Kali Mur.—When caused by taking cold; blood dark, blackish-red.

Dose: During severe acute attacks, five celloids every 15 to 30 minutes, less frequently after relief has been obtained.

SUGGESTIONS

Hot applications, hot pads or cloths wrung out of hot water applied to the lower abdomen may give relief. In severe cases a warm sitz bath for about 10 minutes is helpful. Care should be taken to guard against chilling. Outdoor life and exercise, a nourishing but easily digestible diet, and other means of building up good general health are helpful to prevent the recurrence of dysmenorrhea.

CLINICAL REPORTS

Miss B., age 20, had been suffering for the last two years with severe ovaralgia at the menstrual period. She had been under the treatment of several physicians, and the only relief that they were able to give her was by the use of morphine, that being only temporary. Was called late one night to see her and found her suffering with severe pain in the left ovary of a dull, dragging character, and but slightly intermittent. Patient hysterical and very excitable. Gave her *Kali Phos.* in hot water, every ten minutes for half an hour, when patient went to sleep, not waking until morning, when she was free from pain. Gave *Kali Phos.* once a day for a month, and now, after 18 months, has had no more pain and is feeling better every way.

(G. H. M., M.D.)

Was called to see Miss W., who was suffering from menstrual colic. Gave her *Magnesia Phos.* dissolved in hot water. Three doses relieved her of pain entirely.

(Dr. Thos. A. B.)

Patient had very severe shooting pains during the menstrual periods. The pains were in the stomach and lasted a day or two, commencing in the back and coming directly around and centering in the pit of the stomach. Heat and pressure gave relief. *Magnesia Phos.* given in hot water relieved the pains. She took a few months' further treatment with *Ferrum Phos.*, and the cure was permanent.

(L. T. KEEGAN, M.D., Imperial Beach, Cal.)

Mrs. N., age 30. Menses was several days late, attended by great pain of a constrictive nature. Patient had been caught out in the rain and gotten wet a few days before. Was called and found her suffering almost unbearable pain of a labor-like nature, and bursting headache. Gave *Magnesia Phos.* celloids in combination with *Ferrum Phos.* every 30 minutes. She was relieved in less than an hour. Flow started normally and lasted normal length. Have prescribed for hundreds of similar cases with the same satisfactory results.

(A. C. N., M.D.)

DYSPEPSIA

(See Stomach Disorders, p. 220)

EAR—DISEASES OF

CAUSES AND SYMPTOMS

The majority of ear disorders are of catarrhal origin, frequently transmitted from the nose or throat through the Eustachian tubes into the ears. The care of the nose and throat and the maintenance of a normal mineral balance are important features of the proper treatment of ear troubles.

Internal pain (neuralgic) in the ear is an inflammatory or catarrhal disorder of the middle ear, and is aggravated by closure of the Eustachian tubes.

When the inflammation extends from the membranes of the ear a severe painful trouble known as mastoiditis may arise, which should be treated by a competent physician.

Deafness may be due to a form of paralysis of the auditory nerve, but more frequently it is the result of a catarrhal thickening of the tissues of the middle ear.

TREATMENT WITH THE SCHUESSLER REMEDIES

Ferrum Phos.—Diseases of the ear, when inflammatory symptoms are present, with fever, pain, congestion, etc. Earache, with throbbing, burning pain; also sharp, sticking pains, due to inflammation; hot outward applications relieve by counter-irritation. Noises in the ear from congestion. Temporary deafness, with cutting pain.

Kali Mur.—Secondary affections, after *Ferrum Phos.*, or in alternation with it. After inflammations—the membrane thickened, causing deafness. Deafness, caused by swelling of the Eustachian tubes; also with swelling of the glands of the ear; cracking noise in the ear, when blowing the nose or on swallowing. Earache, with swelling of the glands or membranes of the throat or ear; tongue generally coated white; white catarrhal discharge from the ear.

Kali Sulph.—Earache, with thin, yellow, watery matter. Catarrh of the ear, with the above discharges; sharp pains under the ear. Deafness, from swelling of the parts. Note the tongue, which has a yellow, slimy coating.

Calcarea Sulph.—Deafness, with discharge of thick, yellow matter, sometimes mixed with blood.

Silicea.—Foul, mattery discharges from the ear; *Silicea* hastens suppuration, while *Calcarea Sulph.* shortens the time of the discharge. Hearing dull, when there is swelling and catarrh of the Eustachian tubes, with foul, mattery discharges.

Kali Phos.—Deafness, with noises in the ear; with weakness and confusion; from nervous exhaustion. Ulceration of the ear; pus is dirty, foul and offensive odor.

Magnesia Phos.—Earache of a nervous character, or with sharp neuralgic pains in or around the ear.

Calcarea Phos.—Cold feeling of the ears; pains and aches in the bones around the ear. Earache, with swollen glands in anemic or scrofulous children.

Natrum Mur.—Deafness, from swelling of the cavities and associated with watery discharge, roaring in the ears. Salivation, tongue coated with frothy bubbles of saliva.

Natrum Phos.—Outer ear sore and covered with thin, cream-like scabs. One ear red, hot and frequently itchy, with gastric derangement and acid condition of the stomach. Creamy-yellow coating of the tongue.

Dose: In acute, painful conditions, five celloids of the indicated remedy every one-half to one hour. In chronic conditions, five celloids three or four times daily.

SUGGESTIONS

The application of heat by means of a hot water bottle or a radiant lamp is useful for relieving pain. Local, analgesic ear drops may also act beneficially. If pus accumulates behind the drum-head, puncture (performed by a doctor) may be necessary to prevent spread into the mastoid cells behind the ear.

CLINICAL REPORTS

Mrs. I. S., age 49. As the result of a night trip at high speed in an open car this patient gradually lost her hearing. She had no treatment of any kind up to the time that I saw her, a little more than a year later. By this time she was almost totally deaf.

I prescribed *Kali Mur.* and *Kali Phos.*, five celloids alternately four times daily. In six weeks she was greatly improved, not only her hearing but her general health as well, and in three months she was entirely cured.

(R. G. S., M.D.)

Right ear discharging yellow pus for more than one year. Noises in the ear with loud reports like explosions, etc. Patient very sensitive to cold air. Head sweats and feet often sweaty. *Silicea* for four days. Discharge stopped and patient improved very much.

(F. V. B., M.D.)

Mrs. —, age 45. About three years ago began to be troubled with pain and noises in the left ear; aggravated greatly at the time of the menses; the pain severe and neuralgic in character, extending over the left side of the head. The noises seem to get their character from some pronounced sound which is heard, and this persists sometimes for hours. For the last six months there has been no pain on the left side, but deafness is constant. The right side is now beginning to become deaf, but with no pain and no noises. This has been going on, on the right side, for several months. General health excellent, with the exception of redness, fullness and desire to rub and pull the skin about the neck for a few days after the menses, with marked swelling of the glands of the neck at the same time. This has been noticed only during the time that the ears have been troublesome. The fork is heard best on the left side by bone conduction, and best on the right side by air conduction. Meatus dry and tympanic membrane depressed. Eustachian tube on the left almost occluded, on the right more free. Frequent burning of the auricle on the left side. *Kali Mur.* relieved.

(H. B. B., M.D.)

Patient came from Dallas, had left ear discharging pus for two weeks, also large swelling in front of ear. Had not lain down for two weeks at night, owing to pain. *Kali Mur.* and *Magnesia Phos.* relieved him in a week, then followed with *Silicea,* which brought about a rapid cure.

(Dr. F. V. B.)

Johnnie M., 7 years old. When between four and five years old little Johnnie had measles and about the time they thought him well he took fresh cold. It settled in throat and middle ear. A physician treated him quite a while and had gotten him over the acute part of it, but little Johnnie wound up with chronic affection of middle ear. It ran a yellowish stinking exudation which would excoriate the skin on outside and had formed a large crusty hard scab all around the ear. And the worst of all, both ears were affected.

The boy had been in this condition for more than two years when his father brought him to me with this history and also a run-down system in general. I put the patient on *Ferrum Phos* four times a day at

7 A. M., 11 A. M., 3 P. M., and 7 P. M., also *Calcarea Phos.* four times a day at 6 A. M., 10 A. M., 2 P. M., and 6 P. M. Also irrigated the ear night and morning with normal solution of boric acid first ten days. Then I began to irrigate ear with equal parts of *Silicea* and *Calcarea Fluor.*, 5 grain in ounce of warm water night and morning. I kept this line of treatment up according to the indicated remedies for about eight months. The last 60 days on retiring dropped in each ear warm Mullein Oil (Luyties) and the child made an uneventful and complete cure. He is now a grown young man and can hear as acutely as anyone, and I know it is a fact that the Schuessler Remedies were the cause of the cure.

(Dr. T. W. L.)

I gave a patient *Kali Phos.* for rumbling in the ears, a symptom which was driving the man to distraction. He was also afflicted with a chronic pain in the limbs which was of five years' duration. *Kali Phos.* ultimately brought about a cure.

(H. L. F., M.D.)

Boy twelve years of age had a chronic mastoid infection which had resisted any sort of treatment for a long time. *Silicea* cured the case.

(M. A. T., M.D.)

ECZEMA

CAUSES AND SYMPTOMS

Eczema is a non-contagious, inflammatory skin disease. Probably more than one-half of all skin eruptions are eczematous in nature. Errors in diet and systemic poisoning associated with a disturbance of the balance of the inorganic salts in the body constitute the outstanding causes of eczema.

Eczema in its early stage can be recognized by the appearance on any part of the body of small vesicles which mature, rupture and discharge a serum, which as it dries forms dry scales or small crusts. Some forms of eczema are dry and others continually ooze or weep under the crusts. The skin between the vesicles is inflamed, red and itching (or burning sensation). Pus formation and discharge occurs only from a secondary infection and through neglect of proper measures.

TREATMENT WITH THE SCHUESSLER REMEDIES

Ferrum Phos.—In the beginning for the inflammation, redness, heat and swelling of the skin.

Kali Mur.—Small, white, dry scales on the skin; white-coated tongue.

Kali Phos.—Scabs with offensive, irritating secretions, causing soreness and rawness of the parts, offensive odor. Eczema in nervous persons.

Kali Sulph.—Discharges from eruptions of thin, yellow matter. Eruption suddenly suppressed, with dry skin.

Natrum Mur.—Watery eruption, small vesicles with intense itching. Skin dry and cracked.

Natrum Phos.—Eczema with excessive acid conditions. Creamy, honey colored discharges, golden-yellow crusts.

Dose: Five celloids every one to two hours in acute condition, three or four times daily in chronic eczema.

SUGGESTIONS

Care must be taken to keep the skin clean, yet soap may have to be avoided because it aggravates the condition in many cases. A bland, non-irritating ointment applied to the parts will reduce inflammation and itching. The diet must be carefully studied for the purpose of eliminating the foods which cause aggravations. The movements of the bowel should be kept regular. The kidney functions can be promoted by drinking an abundance of water.

ENURESIS

(Bed Wetting)

CAUSES AND SYMPTOMS

Enuresis, or an involuntary escape of urine, is a condition most frequently found in children and old people. In children enuresis is frequently a habit due to lack of training, in some cases a weakness of the sphincter muscle accompanied by mineral salt deficiencies, is the causative factor.

Excessive acidity, inflammation of the urinary tract, weakness of the muscles of the bladder, are other possible causes, all traceable to a mineral deficiency.

TREATMENT WITH THE SCHUESSLER REMEDIES

Ferrum Phos.—Enuresis from weakness of the sphincter muscles. Also from irritation of the urinary tract; from colds.

Kali Phos.—When the disorder is caused by or associated with nervous weakness.

Natrum Phos.—Excessive acid condition. Also from irritation caused by worms.

Calcarea Phos.—Enuresis in young children and old people due to general weakness, debility or anemia.

Dose for children: Three celloids four times daily. If more than one remedy is required give them in alternation.

SUGGESTIONS

The general hygienic treatment is important. The diet must be simple, and in the case of children a light, early supper is recommended. Beverages of every kind should be restricted after the middle of the afternoon. Cold bathing, exercise and fresh air are beneficial. The child should urinate before retiring and taken up at once if awake during the night and early in the morning before the bladder is full. Thus by breaking the habit the child will often gain control. Punishment for bed wetting is never recommended as this only serves to add to the nervous instability of the child and aggravates the trouble.

CLINICAL REPORT

Recently I have had several cases of nocturnal enuresis in children about five years of age that have apparently been permanently cured by the administration of *Kali Phos.* and *Natrum Phos.* for a few weeks.

The dosage used was four tablets every three hours, and it is astonishing how rapidly these cases improved.

The fact that these remedies may be used with any other form of treatment or kind of medication is a quality that must not be overlooked, and it makes their usefulness cover a very wide range.

If one will but give these remedies a fair trial their use will not be abandoned and the results obtained will be most gratifying.

(Dr. R. G. S.)

EPILEPSY

The characteristic attacks of epilepsy, with their sudden, more or less violent fits and convulsions, make the diagnosis of this disease easy.

The treatment, however, is a task which must be left entirely to a competent physician, who will instruct the family members on the subject of emergency measures needed during or immediately after the attacks.

Therefore, no recommendations on the home treatment of this serious ailment are given here. The following few suggestions are intended only for prevention of injury to the patient during the attacks:

Loosen the clothing and, if possible, place a cork or a piece of wood between the teeth to prevent biting of the tongue. When the convulsions cease draw the tongue forward to prevent choking.

After subsidence of the spasm the patient should be aroused to complete consciousness by giving stimulants (strong coffee) and only then be left to sleep, this will prevent the wretched feeling which follows attacks.

ERYSIPELAS

CAUSES AND SYMPTOMS

Erysipelas is due to an infection by a germ or streptococcus gaining entrance through an abrasion of the skin and finding suitable breeding ground in tissues deficient in the vital mineral elements, and in persons where the general vitality and disease resisting powers are low. It is a highly contagious disease and in the severe form a dangerous one, and whenever possible the treatment should take place under the supervision of a physician. The treatment and suggestions mentioned here are intended for emergency cases only.

Erysipelas usually begins with a chill, vomiting, headache and a rapid rise of temperature. At the point of infection the skin is red and swollen, rising above the surrounding skin with a sharply defined edge. Vesicles form on the inflamed surface, and in severe cases pus and abscesses form. The skin burns and smarts and is fiery hot, the temperature is high, frequently 104 to 106 degrees. This disease becomes especially serious when the eruption appears on the face or head.

TREATMENT WITH THE SCHUESSLER REMEDIES

Ferrum Phos.—Is the principal remedy in the inflammatory stage, fever, flushed face, rapid pulse, vomiting, headache. Rose erysipelas.

Natrum Sulph.—Vomiting of bile. When the skin is smooth, red, shiny, painful and swelled.

Kali Mur.—For the vesicular (blister) form, swelling. Alternate with *Ferrum Phos.*

Kali Sulph.—In the blistering form of erysipelas, to aid in throwing off the vesicles.

Dose: Five celloids every one-half to one hour, less frequently after the fever subsides.

SUGGESTIONS

Rest in bed is required as long as there is fever. Cold sponging or ice packs applied to the head give relief from severe headache. Excessive and painful swelling may be reduced by compresses wrung out in a saturated solution of epsom salts.

The diet should be very light in the liquid form, in severe cases confined to skimmed milk and in all cases an abundance of water.

The patient should be isolated as much as possible from contact with others, and all clothing and bed linen should be boiled.

CLINICAL REPORTS

Mrs. E. A. McD., age 70, tongue thickly coated white, fever 101, constipated, eyes swollen nearly shut, ears swollen three times natural thickness, and auditory canal also closed. Large watery blisters in ears and face were spreading into hair on head, but checked by giving five celloids each of *Ferrum Phos.* and *Natrum Sulph.* in alternation, with hot water, every 30 minutes. As soon as the fever subsided gave *Kali Mur.* instead of *Ferrum Phos.* I allowed no food for three days and had nurse give enema night and morning. A change began for the better immediately as blisters and swelling began to subside after first day. Patient was well and attending to her household duties in one week, which I think good for a woman of her age.

(Dr. M. H. E.)

EYE DISEASES

CAUSES

Mineral salt deficiencies and various factors and influences which cause disturbances in other organs and tissues of the body may also affect the eyes.

The sensitiveness of the eyes and tissues surrounding them and the dangers of injury to this important organ demand proper treatment of even slight disturbances.

Defective vision and serious eye troubles should be treated by a physician or an eye specialist.

TREATMENT WITH THE SCHUESSLER REMEDIES

Ferrum Phos.—First stage of eye inflammations, for the redness, pain, etc. Burning in the eyes; pain in the eyeballs through overstraining the eyes; cold applications relieve. In granulated eyelids, for the pain and inflammation. Eyes blood-shot.

Kali Mur.—Second stage of inflammations, with discharge of white or grayish-white matter. Sore eyes, with specks of white matter on the lids. Granulated eyelids, with feeling of sand in the eyes. Alternate with *Ferrum Phos.*

Kali Sulph.—Late stage of inflammations, with discharge of yellow or greenish matter; yellow crusts on the lids.

Silicea.—Inflammations, with thick, yellow, mattery discharges; sties on the eyelids; little boils and tumors around the eyelids.

Natrum Phos.—Inflammations, with discharges of golden-yellow, creamy matter; eyelids are stuck together in the morning; creamy coating on root of tongue. Squinting when worms are present.

Natrum Mur.—Eye affections, when there is a discharge of watery mucus or flow of tears; discharges cause soreness of the skin or the eruption of small blisters. Granulated eyelids, intercurrently with *Fer-*

rum Phos. and *Kali Mur.* Neuralgic pains, with flow of tears.

Kali Phos.—Weak eye-sight from weakness or exhaustion after disease. Drooping of the lids or squinting after acute ills, from weakness of the muscles.

Magnesia Phos.—Drooping of the eyelids, alternate *Kali Phos.* Contracted pupils; sensitiveness to light; affected vision; sees sparks, flashes and colors before the eyes. Dullness of sight from weakness of the optic nerve. Neuralgic pains; relieved by warmth; spasmodic squinting and twitching of the eyelids.

Calcarea Phos.—Eye affections in anemic patients; intercurrently with other remedies.

Dose: Five celloids every one to two hours in acute conditions, three or four times daily in chronic eye disorders. If more than one remedy is indicated, take them in alternation.

SUGGESTIONS

It is important to prevent infection of the eyes through contact with the hands. Before bringing the hands in contact with the eyes, especially when there is inflammation, the hands should be carefully washed in an antiseptic solution. If any inflammation of the eyes is present a solution of boric acid should be used to bathe the eyes. A few drops of such a solution may also be dropped into the eyes frequently.

Foreign bodies may be removed by rolling the lid outward over a matchstick and mopping the foreign body away with a wisp of cotton, moistened in Colina Eye Lotion (or boric acid solution).

FAINTING

CAUSES

Fainting may arise from sudden change from cold to heat, breathing impure air, fatigue, long fasting, grief, fright, loss of blood, deranged digestion, mental emotions. Nerve fluid contains a large percentage of Potassium Phosphate, and so also does the gray matter of the brain. All of the above symptoms tend to rapidly exhaust nerve fluid, therefore *Kali Phos.* is the remedy to restore the equilibrium.

SUGGESTIONS

When fainting occurs, loosen the clothing about the neck, chest and waist, and move the patient into the fresh air; also place the patient in a reclining position with the head low. Unless the faint be complicated with other serious symptoms, no alarm need be felt, as restoration will, in a short time, be accomplished without interference. If, however, the coma lasts longer than is thought advisable, a little cold water may be sprinkled on the face and neck; should this prove insufficient, a little spirits of camphor or ammonia may be applied to the nostrils for inhalation.

FALLING OF THE WOMB

(Prolapsus)

CAUSES AND SYMPTOMS

Falling of the womb, or prolapsus, is due principally to a relaxation of the muscles which hold the womb in position, arising from a deficiency of the cell-salt *Calcarea Fluor.* Should it tip forward or backward, it is termed anteversion or retroversion, but the treatment is the same in either case. Among the exciting causes of this disease may be noted: straining, lifting, carrying heavy weights, climbing hills or stairs, dancing, etc. It is often the case, also, that general debility, irregular habits, painful menstruation, improper diet, etc., are closely associated with and give rise to prolapsus.

The primary symptoms are: dragging, aching pain low down in the back, just above the hips; dragging pains in the groins; sense of weight in the pelvic region, etc. Secondary symptoms frequently seen are: neuralgic pains in the region of the heart; fainting, pains in the limbs; leucorrhea; painful menstruation; flooding; constipation, etc.

TREATMENT WITH THE SCHUESSLER REMEDIES

Calcarea Fluor.—Is the chief remedy, to tone up the relaxed tissues.

Kali Phos.—When there are nervous symptoms, fainting, "lump" in the throat, suffocative feeling, etc. Alternate with *Calcarea Fluor.*

Calcarea Phos.—Should be given (five celloids three times daily) in all cases of falling of the womb, as a general tonic.

Natrum Mur.—Prolapsus, with weakness, sinking sensations, watery leucorrhea, etc. Must sit down to obtain relief. Alternate with *Calcarea Fluor.*

Ferrum Phos.—Should any inflammatory conditions arise, they will need *Ferrum Phos.* intercurrently.

Dose: Five celloids four times daily.

SUGGESTIONS

The use of the Schuessler Remedies should prove beneficial, especially in cases of comparatively recent origin, and where the elasticity of the muscles which hold the womb in position has not been entirely destroyed. In some cases mechanical or surgical measures may be required, which, of course, require the services of a competent physician.

FELONS

(See Boils, Carbuncles, Abscess)

FEEDING OF INFANTS

Breast milk is a highly specialized food which cannot be exactly duplicated. Breast feeding guards the infant against a variety of ailments to which the bottle fed babies are subject and is a valuable aid to the promotion of health. The death rate among artificially fed babies is many times as great as among breast fed infants and this fact should serve to lead every mother to feed her babe at the breast for at least six months whenever possible. If the period for weaning falls at the beginning of hot weather the babe should be taken off the breast late in the spring or carried over on the breast till cooler days in the fall.

Regularity in feeding time is very important whether the babe be breast or bottle fed and efforts should be made to assure this early in the infant's life. The babe should be wakened at regular intervals during the day for feeding and will quickly establish regularity as a habit. Feeding the infant every time it cries is an exceedingly pernicious habit and one which tends to digestive disturbances as well as a nervous, irritable baby.

If the breast milk disagrees with the baby (which is rare) or if it is not possible to nurse the baby for any reason, we must attempt to supply a food as nearly as possible like the mother's breast milk. This is usually accomplished by diluting top-half cow's milk to about one-fourth strength and adding about one ounce of sugar to the total food for one day. This sugar is best in the form of milk-sugar or corn syrup. The new-born babe will take from one and one-half to two ounces at a feeding at two-hour intervals during the day and once or twice during the night. As the baby ages the interval may be increased as well as the strength of the food. As a general rule the night feeding may be discontinued after the third month.

Orange juice should be given bottle babies and may be given breast-fed ones as well, one to two teaspoonfuls a day after the first month. Thin cereal gruels may be added in the fifth or sixth month, depending on the growth of the baby; vegetables, strained or mashed fine in the seventh or eighth month, and stewed fruits and an occasional egg in the tenth month. Milk, however, should still be given each day and continued throughout childhood, as without milk in the diet it is difficult to keep the necessary supply of vitamins and mineral salts up to the requirements. (See also "Teething Disorders.")

FEEDING OF CHILDREN

Diet of children is an important part of their care and much may be done by careful attention to their food to prevent illness and to promote vigorous, healthy growth. A well balanced diet is necessary at all times but especially in childhood and there are certain requirements in the food of children which are not present later in life.

Children require vitamins and certain of the mineral salts, notably iron, calcium and phosphorus, these being required to build new tissues and for the proper strength of the bony framework, for the formation of teeth of good quality and for the maintenance of normal growth. These minerals and vitamins are required in double or treble the amount of the daily intake of the adult, hence special attention should be paid to the food of the child that these elements are present in sufficient amount every day.

The foodstuffs containing the vitamins and mineral salts in the greatest quantity and most usable form are the green vegetables, eggs, and milk. Milk should constitute a daily article of diet for every child and at least a quart a day should be taken. During the period of active growth and the formation of teeth, lettuce, spinach, carrots and similar vegetables should appear frequently in the dietary with eggs added occasionally but not so often as to allow the child to tire of them.

For older children salads comprise a ready method of adding to the green food supply and should be served frequently. If the child is very active, carbohydrates are necessary to keep up the body heat and sugar is the ideal for this purpose. Pure sweets if not taken in excess nor sufficiently to prevent the child eating sufficient other food to maintain a balance, are proper and necessary for the growing child.

FEVER (SIMPLE)

CAUSES AND SYMPTOMS

Fever is an elevation of the body temperature either local or general. It arises as a result of increased circulation of the blood following a deficiency of the mineral salt *Ferrum Phos.* (iron) in the blood or the tissues of the body. Iron is the chief carrier of oxygen in the body and when deficient, lowered oxygenation results. To compensate for the lower oxygen carrying power of the blood, the circulation is speeded up in order to carry more blood to the tissues, the pulse rate increases, and the symptom called "fever" is produced.

Fever is usually present in the first stage of any disease in which there is inflammation in any part of the body, the intensity of the fever being an indicator of the severity of the disturbance. A clinical fever thermometer should be in every household; it gives information on the patient's condition.

TREATMENT WITH THE SCHUESSLER REMEDIES

Ferrum Phos.—The first remedy indicated in all types of fever. Rapid pulse, flushed face, sometimes accompanied by chilly sensation, vomiting of undigested food. *Ferrum Phos.* should be continued as long as the fever lasts, but to be alternated with such other cell-salts as the subsequent symptoms demand.

In the beginning give five celloids of *Ferrum Phos.* every one-half to one hour, increasing the intervals as the temperature subsides.

Kali Mur.—Second stage of fevers, tongue covered with grayish-white coating. There is usually constipation present with light colored stools. The affected area is congested and swollen.

Kali Phos.—Fevers associated with nervous conditions, excitement and general weakness. Alternate with *Ferrum Phos.*

Kali Sulph.—Evening rise of temperature with hot, dry skin: to promote perspiration.

Natrum Mur.—Early stages, excessive thirst, water does not relieve thirst; dry skn. (Alternate with *Ferrum Phos.*)

Natrum Phos.—Fever with acid conditions, sour vomiting, sour perspiration, bilious condition. Alternate with *Ferrum Phos.*

Dose: Five celloids every one-half to one hour, less frequently as the fever subsides.

SUGGESTIONS

Inasmuch as fever is usually only a symptom of an illness, the general treatment will vary somewhat according to the nature of the illness, and more explicit suggestions are given under the respective headings of diseases associated with fever.

A few general recommendations are applicable to the care of a patient with fever when associated with ailments of practically every description, namely:

The patient should remain in bed, quiet, as long as there is an active fever. The diet should be light, in the liquid form, and easy to digest. The intestinal tract should be kept clear of foreign material. In some cases the use of a mild laxative remedy is desirable in other cases an enema containing a little salt is preferable.

A simple measure to reduce high fever is the hot sponge bath. Sponge the body lightly with a cloth wrung from hot water, then dry off gently, or allow the moisture to evaporate, thus cooling the surface. A cool sponging is more effective in some cases. In case of high fever and throbbing head, a cloth wrung from cold or iced water may be placed on the head.

GALL STONES

CAUSES AND SYMPTOMS

Catarrhal inflammation of the gall-bladder, with a deficiency of mineral salts, especially of *Natrum Sulph.* and *Natrum Phos.* affect the fluidity and viscosity of the bile, cause cholesterin deposits and eventually the formation of gall stones. Prevention of the thickening of the bile is the surest way of stopping the formation of stones.

The existence of stones in the gall-bladder does not always produce painful symptoms, but if the stones become impacted in the narrow ducts, intense colic, vomiting and even collapse results, and removal may require an operation. If the stones are not too large they may find their way through the ducts into the intestinal tract and cause no further trouble.

TREATMENT WITH THE SCHUESSLER REMEDIES

Natrum Sulph.—Biliousness, bitter taste with grayish-green coating of the tongue, vomiting. Should be used periodically by bilious persons to prevent the formation of gall stones.

Calcarea Phos.—Also used to prevent the formation of gall stones by equalizing the balance of this salt in the system. (Use in alternation with *Natrum Sulph.*)

Natrum Phos.—Acid conditions of the stomach; creamy, yellow coating of the tongue, jaundice with pain in the region of the liver.

Magnesia Phos.—Gall stone colic, severe pain. (Take five celloids every 10 to 15 minutes.)

Dose: Five celloids every two hours. For the preventive treatment, five celloids three times daily.

SUGGESTIONS

Gall stone colic can be relieved by hot applications, by an electric pad or hot water bottles. Drinking of hot water helps to move along accumulated gas, and to some extent stimulates the passage of the stone along the biliary ducts.

Those inclined to biliousness, liver and gall-bladder disorders, should also regulate their diet, avoid fats, fried food, cheese, cream, egg-yolk, sweetbreads, liver. Green vegetables, fruits, roasted lean meat and fish are allowed.

Fresh air and exercise tend to promote combustion in the body and prevent stagnation of the bile. Stoutness after middle life is a contributing factor to disturbances of the function of the liver and gall-bladder.

The symptoms of gall stones in some cases become very severe and special palliative measures under the direction of a physician may be required. When the attacks occur frequently, despite medicinal treatment, and in cases where the obstruction of the gall-bladder duct is very persistent, surgical treatment may become necessary.

GLANDS, DISEASES OF

CAUSES AND SYMPTOMS

The glands are organs of various sizes found in different parts throughout the body. Some of them secrete substances which are used within the system, and are indispensable to the life-functions of the body. Other glands serve for the elimination of undesirable matter from the body.

Every gland must function properly in order to maintain a state of good health. Mineral salt deficiency is a leading cause of disturbances in these functions, and unless corrected may bring about glandular diseases of serious nature.

The most prominent symptom of glandular disorders is the swelling or enlargement of the gland or glands. This condition can generally be overcome through the use of the indicated deficient mineral salts. If within a reasonable time amelioration has not been obtained, the consultation of a competent physician is urged, as additional measures may be required for the correction of the trouble.

TREATMENT WITH THE SCHUESSLER REMEDIES

Kali Mur.—Is the principal remedy in glandular swellings, when the gland is not of stony hardness. Swelling of the glands of the neck and throat. Enlargement of glands.

Ferrum Phos.—For the fever and pain in acute swelling of the glands.

Calcarea Fluor.—Swelling of glands, when of stony hardness. Chronic, very hard inflammatory conditions of the glands.

Natrum Mur.—Swelling of the glands, with watery symptoms, excessive secretions of saliva, etc. Alternate with *Calcarea Fluor* or *Kali Mur.*

Calcarea Phos.—Chronic enlargement of the glands. (Alternate with *Kali Mur.*). Simple goiter.

Silicea.—Scrofulous glands, alternate with *Calcarea Phos.* Swollen glands which are inclined to suppurate.

Calcarea Sulph.—When the glands are discharging pus, to control suppuration.

Dose: Five celloids every two hours in acute glandular disorders; four times daily in chronic conditions.

SUGGESTIONS

Local applications are generally of only slight value and in some cases only tend to increase the irritation. Proper, adequate, well balanced nourishment, fresh air and sunshine and regular bowel elimination are hygienic measures which will prove helpful in correcting glandular disorders.

CLINICAL REPORTS

A young child about three years of age had a swollen gland just below the ear as large as the half of a big orange. The parents refused operation and wished that

I try to reduce it with other treatment. *Calcarea Fluor.* cured the case completely.

(M. A. T., M.D.)

GOITER

CAUSES AND SYMPTOMS

Goiter is an enlargement of the thyroid gland, which is located on the front of the wind-pipe above the collar bone. The same causes give rise to this condition as mentioned in the preceding article "Diseases of the Glands." Aside from the usual mineral deficiencies, especially lime, or *Calcarea Phos.*, the lack of iodium in the body is a contributing cause to goiter.

TREATMENT WITH THE SCHUESSLER REMEDIES

Calcarea Phos.—Is the chief remedy in simple goiter and similar glandular swellings; to be used in alternation with *Natrum Phos.*, when acid symptoms are present. Goiter in anemic patients.

Calcarea Fluor.—If the gland is of stony hardness.

Natrum Mur.—Goiter, when accompanied with much secretion of saliva.

Natrum Phos.—When there are acid symptoms present. Give in alternation with *Calcarea Phos.* or *Calcarea Fluor.*

Dose: Five celloids four times daily.

SUGGESTIONS

In simple cases of goiter the persistent use of the indicated remedy over a period of several months should bring satisfactory results.

A certain form of malignant goiter, known as exophthalmic goiter, is revealed by a protrusion of the eyeballs, nervous trembling, and rapid action of the heart. This disease may prove dangerous to life and should be treated under the direction of a physician.

CLINICAL REPORTS

Mr. E., age 17, of Bellaire, Kans., came to me with a goiter from which he had been suffering for a period of time.

I placed him on *Calcarea Fluor.*, *Natrum Mur.* and *Kali Phos.* to be taken in doses of four celloids, alternating them four times a day. After two months' time the enlargement of the thyroid gland had entirely disappeared. No recurrence of this case up to the present time.

(Dr. Y. N.)

GONORRHEA

This inflammatory disease of the genito-urinary tract is caused by infection with the gonococcus.

No attempt should be made to treat this disease without the services of a competent physician, because of the possible damaging results of faulty procedure in the treatment.

The use of the Schuessler Remedies, especially of *Kali Mur.* and *Ferrum Phos.*, is beneficial, but the prompt and complete elimination of the effects of the gonococcus demands local treatment, the character of which in each individual case must be left to the judgment of the physician.

GOUT

CAUSES AND SYMPTOMS

Gout is a condition similar in its manifestations to rheumatism. It usually comes on suddenly with pain in the great toe joints. The affected joint becomes red and swollen and extremely painful. The attacks subside and return again a few days later. When the trouble assumes the chronic form the joint becomes permanently enlarged and remains tender and sensitive at all times.

Frequently an increased acidity in the body can be noted prior to an attack of gout. Persons who lead a sedentary life, indulge in rich food, appear to be predisposed to this disease. Heredity also is an important factor in gout, and statistics show that it runs through families for many generations.

TREATMENT WITH THE SCHUESSLER REMEDIES

Ferrum Phos.—For the fever and other inflammatory symptoms.

Natrum Sulph.—The chief remedy in this affection, especially if due to high living, or when there are bilious symptoms present. In acute attacks alternate with *Ferrum Phos.*

Natrum Phos.—If acid conditions are in evidence, cream-coated tongue, sour-smelling sweat, etc. Chronic gout.

Dose: Five celloids every one to two hours during acute attacks, four times daily in chronic gout.

SUGGESTIONS

The regulation of the diet is important. Plain, simple foods, principally vegetables, are recommended. Alcoholic drinks especially cause aggravations. The local application of an analgesic balm to be covered with cotton held in place with a bandage, is helpful to relieve attacks. Between attacks the patient should indulge in mild but regular exercise.

HAY FEVER

CAUSES

Hay fever is a recurrent affection of the respiratory tract, periodic in appearance. Inhalation of the pollen from various plants, particularly from the rag-weeds, has been regarded as the cause of this trouble. Schuessler considers that the deficiency of certain mineral salts, especially that of *Ferrum Phos.*, *Kali Mur.* and *Natrum Mur.* is the underlying cause of hay fever, and the pollen only a contributory cause.

The system, whenever the mineral salts are properly balanced, he contends, is immune to the effects of the pollen. It appears, therefore, that the rational and most successful treatment of hay fever will be found in the preventive treatment, by means of correcting the deficiencies of the salts mentioned above.

It is suggested that patients who are annually troubled with hay fever should use the respective Schuessler Remedies approximately four to six weeks prior to the expected attack of hay fever, and continue the treatment during the actual hay fever season.

TREATMENT WITH THE SCHUESSLER REMEDIES

Ferrum Phos.—Severe acute attacks, membrane congested, inflamed, burning sensation, headaches.

Natrum Mur.—Copious, thin watery discharges, sneezing, watery discharge from eyes.

Kali Mur.—Advanced stage, discharge of white phlegm, catarrhal conditions.

Dose: Five celloids every one-half hour during acute attacks, less frequently as the condition improves.

SUGGESTIONS

A regulation of the diet, that is a reduction in quantity of meats, sweets and starches has helped to modify the severity of the symptoms. Preventive measures as suggested in this article are desirable. In addition to the use of the Schuessler Remedies, frequent cleansing or sterilization of the nasal cavities with an antiseptic solution is recommended.

CLINICAL REPORT

Mr. W., age 55. Severe hay fever annually from first flowers until frost. After I gave him *Natrum Mur.* and *Ferrum Phos.* every hour, and one dose per day of *Calcarea Phos.* and placed him on fruit diet, party did not have hay fever during summer.

(Dr. F. H.)

HEADACHE

CAUSES

A headache, in most cases, is only a symptom of a disorder. It may be associated with a disturbance in almost any part of the body. Some of the most prevalent underlying conditions are: Head colds, catarrh, indigestion, biliousness, nervous disorders, local infections, constipation, high blood pressure, weak eyes, etc.

In order to obtain lasting benefits from the treatment of headaches it will be necessary to analyze the source of the trouble very carefully, and begin with the elimination of the basic disturbance and the correction of the mineral salt deficiencies associated therewith.

Sedatives, such as aspirin and similar drugs, have the power to suppress the pains temporarily, but they have no true curative influence upon the underlying condition.

TREATMENT WITH THE SCHUESSLER REMEDIES

Ferrum Phos.—Headaches of an inflammatory nature, with bruising, pressing, stitching pains; congestive headaches, red face or blood-shot eyes; throbbing, beating headache, in the temples or over the eye; blind, sick headache, with vomiting of undigested food. Scalp sore or tender to the touch; headaches from excessive heat or cold. Congestive headache at the menstrual period. Alternate with the remedy indicated by the color of the tongue. All pains worse from motion and noise.

Kali Phos.—Headaches of a purely nervous character, resulting from overstrain of the mental faculties, worry, sleeplessness, etc. Headaches of nervous, sensitive, pale, irritable or excitable patients. Student's headaches; headaches, with inability for thought, better under cheerful excitement, or with gentle motion; with empty feeling at pit of stomach; pains at the back of the head, with weariness or exhaustion. Tongue is frequently coated like stale mustard; bad breath.

Kali Mur.—With sluggish action of the liver; white-coated tongue or vomiting and hawking of milk-white mucus.

Natrum Mur.—Dull, heavy headaches, with profusion of tears, watery discharge from the nose, or excessive flow of saliva; frequently associated with constipation of the bowels. Headaches, with sleepiness, unrefreshing sleep. Vomiting of watery, transparent fluids. Headaches of young girls with irregular menstruation, with watery symptoms. Pains are generally worse in the morning and disappear in the evening.

Magnesia Phos.—Neuralgic headaches, pains are excruciating, stinging, shooting, darting, intermittent or in paroxysms. Heat relieves, cold aggravates; headaches with "sparks" before the eyes. Nervous headaches, with crampy pains, worse from cold draughts of air.

Natrum Sulph.—Bilious headaches; vomiting of bile, bitter taste, greenish-gray coated tongue, colicky pains or bilious diarrhea. Sick headaches, with giddiness, vertigo and dullness. Violent pains at the base of the brain or on top of the head, cannot tolerate noise.

Natrum Phos.—Headache, with acid conditions; on the crown of the head; on awaking in the morning; with acid, sour risings; vomiting of sour or acid fluids. The tongue generally has a creamy, moist, yellow coating on the back part or in the roof of the mouth.

Calcarea Phos.—Headaches, with cold feelings in the head, or with creeping coldness and numbness on the head. Pains are worse from heat or cold.

Kali Sulph.—Headaches which grow worse in the evening or in a heated room, better in the cool open air.

Dose: Five celloids of the indicated remedy (or remedies) every one-half to one hour during acute attacks, at longer intervals after relief is obtained. When more than one remedy is required use them in alternation.

SUGGESTIONS

The general treatment as suggested for the various disorders associated with headaches should be observed. Some forms of headaches can be relieved by hot applications to the head, others respond better to cold ones. A hot foot bath with mustard usually brings relief from congestive headaches. While taking a hot foot bath a cold cloth or an ice bag may be applied to the head.

Regularity in bowel elimination will correct certain types of headaches. A mild laxative remedy is required in many cases.

The local application of Menthol Balm to the forehead and temples will bring prompt relief in certain types of headaches.

CLINICAL REPORT

Miss B., age 19, a teacher, while attending County Normal, became afflicted with "zigzag appearance around all objects" and at times either "double vision" or "seeing only half of an object." She called upon a physician there, but received no help. On Saturday she returned home, unable to continue her work at the Normal. She consulted me. I found that she suffered from headaches that were worse in the morning on waking, and from mental exertion, and was better when sitting or lying still. Here was a chain of symptoms calling for *Natrum Mur.* I prescribed *Natrum Mur.*, five celloids every hour. On Monday morning Miss B. returned to her work at the Normal, well.

(Dr. A. E. W.)

While attending college I was troubled with a very severe headache, due to overstrain of the mind, from too close application to study. The pains, of a neuralgic character, were accompanied by roaring and buzzing in the ears, dimness of vision, specks before the eyes; could not think properly, everything seemed so confused; there was also a nasty, disagreeable, brownish-yellow coating on my tongue. Consulted one of the professors, who prescribed *Kali Phos.* five grains every three hours. Began to get better at once, continued taking the remedy for about three weeks, gradually taking it less frequently, and since then I am troubled with slight headaches only occasionally.

(Dr. H. H. H.)

Girl, age 16, has suffered for years from periodically returning headaches. The pain is concentrated in the right temple, and of a boring nature, as if screw were being driven in—as the patient expresses herself. Preceding this pain there are burning sensations in the pit of the stomach, bitter taste in the mouth and lassitude. These symptoms are only felt at night or in the morning. When the attack comes on the patient is quite unable to attend to ordinary duties. Generally, vomiting of bile follows, and then gradual improvement sets in. The use of *Natrum Sulph.* several times daily has entirely prevented the return of these attacks of headache. (Dr. W. H. S.)

Mrs. H., age 50, has been troubled with headaches for many years. The most severe attacks usually manifested themselves in the morning, immediately on arising. The pains extended from the nape of the neck to the top of the head, and were usually accompanied by nervous spells, trembling of the hands, blurred vision. The pain would gradually lessen in severity and cease entirely by noon. A dietitian made a careful check-up for a possible food intolerance. Regulation of the diet, according to the report of the patient, brought no noticeable change in the frequency and severity of the headaches. The next step the patient took was a visit to the oculist, and the re-examination of the eyes with a change of eye-glasses. This also proved ineffective. X-rays of the head and blood-tests did not disclose the actual causes of the trouble, which appeared to be essentially of nervous origin. About six months ago the patient began the use of *Kali Phos.* in doses of five tablets four times daily. During attacks of headache, the *Kali Phos.* was alternated with a dose of *Magnesia Phos.* in a little hot water every one-half hour. This treatment proved very satisfactory. The attacks of headaches soon became less frequent and less severe. The patient is now troubled with headaches only at long intervals, and then a few doses of *Magnesia Phos.* and *Kali Phos.* will serve to promptly relieve the attack. (C. O. H.)

HEART DISORDERS

CAUSES AND SYMPTOMS

Whenever there are heart symptoms present it is advisable to submit to a careful medical examination for a definite diagnosis of the trouble. Certain organic heart diseases are almost impossible to cure, yet a carefully prescribed mode of living may prolong the patient's life for many years. The treatment of heart diseases, whenever possible, should take place under the personal direction of a physician.

Minor heart symptoms such as occasional palpitation, heartburn, etc., do not always indicate the presence of serious organic heart disease. Quite frequently gas from indigestion and intestinal fermentation produce pressure upon the heart which causes unpleasant heart symptoms.

Should these heart manifestations persist after the digestive disorder has been corrected, it will then be necessary to consult a physician for further treatment.

HEMORRHAGE

A hemorrhage is the escape of blood anywhere in the body from the confines of the blood vessels. No hemorrhage, excepting perhaps an occasional nose-bleed or when caused by an injury, is without significance, and the discovery of the cause and the correction thereof is a task for a physician.

Every person, however, should have some knowledge of the emergency treatment of such conditions regardless of the cause. Some forms of hemorrhage will cease without serious loss of blood if the patient is kept quiet, other cases demand further measures to check the loss of blood.

The use of the Schuessler Remedies, of course, cannot be considered an emergency treatment for hemorrhages, but will serve as a prophylactic measure to prevent those hemorrhages which can be traced to mineral salt deficiencies in the system. The most useful remedies for this purpose are *Ferrum Phos.* and *Kali Mur.*

Immediate measures in case of hemorrhage are absolute rest, pressure on the external sites of bleeding, elevation of the part with head lowered (except in nose-bleed) and the application of ice at the point of hemorrhage.

If the source of hemorrhage is accessible, plugs of gauze or pledgets of cotton soaked with alum water or Hamamelis (witch hazel) extract at the seat of the hemorrhage is advisable.

Hemorrhage of the limbs can be controlled by the use of constricting bandage or elastic rubber straps.

Local treatment of hemorrhages of internal organs is practically confined to the use of ice bags over the involved area.

CLINICAL REPORTS

Patient suffering from hemmorrhage after extraction of tooth. Doctor called to house. Gave ergot. Bleeding persisted. Another doctor called. Gave Ceanothus in teaspoonful doses every hour. Still no results.

I was called in after three days of profuse bleeding. Packed the cavity with cotton, soaked with Ceanothus and gave *Ferrum Phos.*, five-grain celloids, one every hour. Bleeding stopped in a few minutes and no more trouble. (Dr. A. H.)

I have had a large number of rather severe cases of nasal hemorrhage. As an experiment I used *Ferrum Phos.* and the results have been startling. I find that I can stop the hemorrhage in less time and the results are permanent, as none of the cases on which I have used this remedy have had a return.

(H. L. F., M.D.)

For many years I have been troubled very frequently with epistaxis (nose-bleeding). These attacks would set in from blowing of the nose and sometimes after physical exertion.

About a year ago I began using *Ferrum Phos.* and *Kali Mur.* in alternation every two to three hours, continuing the treatment for about three months. Have had no more attacks since then. (L. B. B.)

HEMORRHOIDS

CAUSES AND SYMPTOMS

The underlying pathological factor in origin of hemorrhoids (piles) is a relaxation of the muscular fibers of the rectal veins and consequent dilation, due in a large measure to a mineral salt deficiency, particularly of *Calcarea Fluor.* and *Ferrum Phos.* Mechanical contributing causes are constipation, straining at stool and sedentary occupations requiring sitting for long periods of time, which aid engorgement and dilation of the hemorrhoidal veins.

Usually a distinction is made for various types of hemorrhoids, such as internal and external hemorrhoids, also bleeding and non-bleeding ones. A differentiation of more significance to the patient would be a classification as follows: 1. Hemorrhoids of recent origin with slight or moderate dilation of the rectal veins, amenable to proper medicinal treatment. 2. Hemorrhoids of long standing with extensive chronic dilation of the veins or strangulation, bleeding or oozing of mucus, types which can be relieved by medicinal treatment, but usually demand surgical treatment for complete removal.

TREATMENT WITH THE SCHUESSLER REMEDIES

Calcarea Fluor.—Is the chief remedy in this condition, for the relaxed condition of the muscular fibers. Bleeding, itching piles, with pressure of blood; pain low in the back; piles with chronic constipation.

Ferrum Phos.—For the soreness and inflammation; also when discharging bright-red blood.

Kali Mur.—When the blood is thick and dark. White-coated tongue; inactivity of the liver. Also hemorrhoids with discharge of whitish mucus. Alternate with *Calcarea Fluor.*

Natrum Sulph.—Piles, with bilious conditions; heat in the lower bowel; excess of bile. Alternate with *Calcarea Fluor.*

Magnesia Phos.—Sharp, cutting or stinging pains. Severe, spasmodic attacks.

Natrum Mur.—When the stools are hard, dry and crumbling. Alternate with *Calcarea Fluor.*

Calcarea Phos.—Intercurrently in chronic cases of piles, as a constitutional tonic, especially in anemic persons.

Silicea.—Very painful hemorrhoids, sensitive to touch. Prolapsed, strangulated and suppurating piles. Constipation, hard, dry stools.

Dose: Five celloids of the indicated remedy every one-half to one hour in acute attacks; four times daily in chronic conditions.

SUGGESTIONS

During acute attacks of inflammatory piles rest in bed is required and either cold or hot compresses may give relief, also injections of small quantities of iced water. If irritating discharges are present enemas containing baking soda, boric acid and glycerin in water are recommended. Good astringent salves or rectal suppositories also act beneficially. Constipation, the principal contributory cause of hemorrhoids, must be corrected.

If bleeding becomes persistent a careful examination should be made by a physician.

CLINICAL REPORT

I have in mind a patient who came to me to be treated for itching hemorrhoids of long standing. He said the itching was almost intolerable; said that he had been treated by a number of good physicians with no relief. He was cured by *Calcarea Fluor.* in a few weeks. It is now more than two years, and there has never been any return of the trouble. (Dr. I. M. H.)

HICCOUGH

CAUSES AND TREATMENT

Hiccough is usually the result of disturbances of the stomach as in indigestion. It is also caused by an irritation of the nerves controlling the diaphragm, brought on by excitement or cold drinks. A predisposing cause is a deficiency of mineral salts in the nerve tissues, making these nerves more than normally irritable. It may become a very annoying condition and even dangerous to life when it occurs after operations.

Magnesia Phos. is the principal Schuessler Remedy in hiccough. It should be taken in doses of five celloids every 15 to 30 minutes. A piece of ice dissolved in the mouth or a cold drink sipped slowly will in some cases relieve the spasm. Stretching of the diaphragm muscles by holding a deep breath may stop the hiccough.

In infants subject to hiccough after nursing, lay the baby on the stomach over the lap. This pressure will force the gas from the stomach and usually stops the hiccough.

HIP JOINT DISEASE

A tubercular inflammation of the structures in and about the hip joint, a disease generally confined to children. The onset of hip joint disease is usually gradual, the characteristic limp is often absent until destruction of the joint structure has begun to take place.

This disease is of too serious nature to justify treatment without medical supervision. The replacement of deficient mineral salts is, of course, important, but orthopedic treatment and other general measures are required to stop the progress of this disease.

HOARSENESS

CAUSES

Hoarseness is due to a swelling of the vocal cords during a cold, sore throat, also to unusual strain placed on the cords. Failure to recover the voice promptly after such strains points to deficiency of salts of iron and potash.

TREATMENT WITH THE SCHUESSLER REMEDIES

Ferrum Phos.—Painful hoarseness of speakers and singers, with inflammation; from overexertion of the voice, or from taking cold; scraping of the throat, with sensation of dryness.

Kali Mur.—Hoarseness, and huskiness from cold. Second stage, loss of voice.

Kali Phos.—If there is exhaustion or nervous depression; tired, weary feeling in the throat.

Calcarea Phos.—When the phlegm is albuminous (like white of an egg); alternate with *Ferrum Phos.*

Dose: Five celloids every one-half to one hour during severe attacks, less frequently after the condition has been relieved.

SUGGESTIONS

Avoid irritating food or drink, night air, overexertion of the voice. Local compresses may give relief. At night a local application of Luyties Panacea Ointment covered with a bandage will greatly assist to relieve the congestion.

In case of persistent hoarseness examination of the vocal cords and larynx should be made by a physician.

CLINICAL REPORT

B. J., a minister, came to our town to hold a series of meetings. I noticed he was very hoarse and spoke to him about it. He said he was afraid his voice was going to quit because it had given him a lot of trouble the past year. He had tried quite a number of remedies and consulted quite a few physicians in his travels, but none seemed to give any permanent relief. I gave him quite a few celloids of *Ferrum Phos.* and *Kali Mur.* I told him to begin about 4 o'clock of the afternoon and take one on tongue, let dissolve, and swallow without water. They put a stop to his hoarseness, and he had no trouble during that series of meetings. When he left I gave quite a lot of them to him, enough for two or three months. They completely cured him of his hoarseness. He said he had forgotten he had a throat. So *Ferrum Phos.* and *Kali Mur.* made it possible for him to continue to do his work as a minister.

(DR. T. W. L.)

HYSTERIA

CAUSES AND SYMPTOMS

Hysteria is a condition characterized by more or less frequent attacks of a lack of self-control. There is a great sensitiveness of the nervous system, distress of mind, laughing and crying alternately, there may be convulsive movements, and the attack frequently culminates in unconsciousness. If the condition is permitted to go without restraint it may become a habit. It is a condition most frequently affecting young women.

Undoubtedly mineral salt deficiencies lie back of the improper functioning of the nervous system and of other contributing causes to hysteria, such as disorders of female organs and of the stomach and intestinal tract. Heredity also seems to be a factor.

TREATMENT WITH THE SCHUESSLER REMEDIES

Kali Phos.—Is the principal remedy in this disease. Nervous attacks, from intense emotion, despondency, passion, or any other cause. Feeling of ball in throat; hysterical laughter, crying or screaming.

Natrum Mur.—Alternate with *Ferrum Phos.,* if hysteria is associated with sadness, moody spells or irregular menstruation.

Calcarea Phos.—Intercurrently in all cases where there is a tendency to anemia.

Dose: Five celloids every two hours.

SUGGESTIONS

Proper companionship is quite important, as well directed mental suggestion has a beneficial influence upon the patient. Firmness is needed in dealing with hysterical persons, but they should not be scolded or punished. They should be treated tactfully. As a rule the patient should be kept fairly quiet, free from excitement and worrying; mental rest is more important than physical rest. In fact, mild exercise without tiring the patient is desirable. Often a hot bath will quiet the nerves and stop an attack of hysteria. The food should be nourishing, varied and invitingly prepared and served. It should be remembered that it is very difficult to cure a hysterical temperament, but it can be controlled.

CLINICAL REPORTS

Lady, about 45, very nervous. Came to me for dentistry. She was void of self-control. Fainted in the operating chair before attempting to have dental service rendered and would not allow dentistry to be done for her that day. She took *Kali Phos.* and *Ferrum Phos.*, five celloids alternately for two weeks, and on account of the death of her mother she was delayed one week longer, continuing the celloids. She was so much improved that she was fearless and as easy a dental patient to work for as anyone.

(Dr. E. B. G.)

Miss R., age 16, menstruated once when thirteen years old, and not since. Was a remarkably healthy and well-nourished girl until three months before she consulted me, languid and weak and suffered much with her stomach. When I was called to see her she was not able to retain her food, and it would be vomited as soon as taken. Complained of great pain in the stomach immediately after eating even the lightest food. On several occasions the pain caused severe hysterical convulsions. The tongue was but slightly coated white; bowels constipated; abdomen tympanitic and very sensitive to the slightest pressure; no fever, but much thirst; water, like food, was ejected as soon as swallowed. At first I thought that I had a case of nervous dyspepsia to deal with, but finally concluded that I had a case of true hysteria, as she was so extremely nervous and hyperesthetic all over and much given to tears when anyone was around. I also found that she had the convulsions whenever her plans were thwarted in any way, but upon my threatening to put her in cold water if she had another, she stopped them. *Ferrum Phos.* relieved some of the symptoms in one week, and *Kali Phos,* relieved all the other symptoms in two weeks more, and my patient was soon as strong and healthy as before her illness. Menstruation returned two months after, and she has been all right since.

(G. H. M., M.D.)

INDIGESTION

(See Stomach Disorders, p. 220)

INFLAMMATIONS

CAUSES AND SYMPTOMS

An inflammation is in most instances simply a symptom of a disorder, the contributing causes of which, aside from mineral salt deficiencies, quite frequently are infections by bacteria. Pain, heat, swelling and redness of the part affected are the common characteristics of an inflammation, which when it involves certain organs such as the lungs, kidneys, stomach or intestines may assume a form dangerous to life and should therefore be treated by a physician.

Minor inflammatory conditions can be satisfactorily managed at home, especially with the aid of the Schuessler Remedies which should be selected according to symptoms and conditions and not according to the name given the disease.

TREATMENT WITH THE SCHUESSLER REMEDIES

Ferrum Phos.—The remedy for the initial stages of inflammations, no matter in which part of the body. There is considerable heat, redness and pain, but no swelling of the parts affected, nor are there secretions or exudations from the seat of the inflammation. Also the remedy with the presence of fever, rapid pulse, headache.

Kali Mur.—In the second stage when there is swelling or a whitish exudation, tongue coated white.

Kali Sulph.—In the advanced stages when the discharges have changed to yellow.

Calcarea Sulph.—In the final stage of resolution, the discharges are thick, yellow matter, sometimes streaked with blood.

Kali Phos.—Not especially the remedy for the inflammation, but for nervous symptoms, restlessness and sleeplessness if they arise. Alternate with other indicated remedy.

Calcarea Phos.—After the symptoms of inflammation have subsided, as a tonic especially for anemic patients.

Dose: Five celloids every one-half to one hour in acute conditions, at longer intervals after amelioration. If more than one remedy is used take them in alternation.

SUGGESTIONS

During the active stage of inflammations, excepting during those of very minor nature, the patient should remain at rest, the diet should be light, non-irritating, preferably in liquid form. Hot applications usually bring relief. If there is congestion in the head apply cold compresses to the head. The functions of the bowels should be kept regular and kidney elimination should be promoted by drinking of liberal quantities of pure water.

After recovery, if there remains weakness, debility, the use of a good tonic is recommended.

INFLUENZA

CAUSES AND SYMPTOMS

Influenza is an acute infectious disease, frequently appearing in epidemic form. It is commonly accepted that infection by the bacillus occurs through the respiratory tract. The bacilli lodge in tissues weakened by the lack of normal supply of the indispensable mineral salts, rapidly multiply and cause the illness known as influenza.

The onset of the trouble is usually sudden, beginning with a chill followed by fever, headaches, aching of the muscles, chiefly of the back and limbs, there is nervousness, great weakness and even prostration. If the intestines are involved, diarrhea is also present. During certain epidemics the disease seems to be more virulent and destructive than during others. Much depends also upon the vitality of the patient in resisting the disease and the always present liability to dangerous complications, especially pneumonia. No specific remedy for the prevention or treatment of influenza has thus far been discovered, and the physician must depend essentially upon palliative medication and suitable general measures, such as rest, diet, etc., for the treatment of this disease.

The correction of mineral deficiencies is an important part of the rational treatment for influenza, as it is for other abnormal conditions.

TREATMENT WITH THE SCHUESSLER REMEDIES

Ferrum Phos.—In the first stage with chills, followed by heat, fever, headache, anxiety, vomiting.

Kali Sulph.—In alternation with *Ferrum Phos.* to promote perspiration. Dry skin, feeling of heat.

Kali Mur.—Sore throat, the tongue has a white-coated appearance.

Natrum Mur.—For the watery conditions, tears, running from the nose, sneezing, dry throat, etc. Mental depression. Alternate with *Ferrum Phos.*

Natrum Sulph.—When bilious symptoms are predominant; vomiting of bile, pain in the liver, yellow skin, diarrhea, etc. Alternate with *Ferrum Phos.*

Magnesia Phos.—For the sharp, shooting and darting pains. Neuralgia following influenza.

Calcarea Phos.—Weakness and physical lassitude during and after convalescence.

Dose: Five celloids every one-half to one hour in the early, acute stage. Later five celloids every two hours. In most cases of influenza more than one of the Schuessler Remedies is indicated. Two or more remedies can be taken at the same time; it is preferable, however, to take them in alternation, about one hour apart.

SUGGESTIONS

Rest in bed from the beginning until the fever is gone and the temperature has remained normal for at least 24 hours. The diet should be liquid and return to regular diet should be gradual. Give an abundance of water and hot drinks. Keep the bowels open. Keep the mouth, throat and nasal passages clean by the use of Creozone Antiseptic Solution. If the fever is high a sponge bath will give relief.

If there is weakness after recovery the use of Luyties Tissue Tonic is advisable.

CLINICAL REPORT

Mr. R., age 26, went to work in the morning feeling well; about 10 o'clock he began to experience a very tired, weary feeling, and suddenly became very weak. There was considerable sneezing and much lachrymation, with thin watery discharge from the nose. He was compelled to go home, and then sent for me. The temperature for the following two days was 103° to 104° with little or no further variation. He complained of great soreness of the muscles, severe backache and bone pains, and other characteristic symptoms of influenza. I prescribed the usual remedies but did not get satisfactory results. I then prescribed *Natrum Sulph.* to be given every hour, five 1-grain tablets at a dose. The result was striking. In a few hours he felt so much better, it was with difficulty that he was kept in the house, and he returned to work the following morning, as he felt entirely recovered from the attack of influenza. (H. L. D., M.D.)

JAUNDICE

CAUSES AND SYMPTOMS

There are two leading contributory causes of jaundice: 1. Obstruction of the bile ducts. 2. Disease of the liver and gall-bladder.

The first cause is less serious, as the jaundice disappears as soon as the ducts permit the passage of the bile. Liver diseases, on the other hand, are sometimes difficult to correct. Malignant jaundice from atrophy of the liver is a very serious disease and should not be treated without consultation of a physician.

The outstanding symptom in jaundice is the lemon-yellow tinge of the skin and of the whites of the eyes. The tongue is coated white or yellow, there is a bitter taste in the mouth, the stools are white or light-yellow, the urine is highly colored.

TREATMENT WITH THE SCHUESSLER REMEDIES

Ferrum Phos.—Early stage, inflammatory condition, fever, pain in liver, vomiting of undigested food.

Natrum Sulph.—Congestion of the liver and gall-bladder with resulting jaundice. Biliousness, flatulence, cutting pains, greenish stools.

Kali Mur.—Jaundice with catarrhal condition. Constipation, light colored stools; white coated tongue, bitter taste. Vomiting of thick white mucus.

Natrum Mur.—Jaundice associated with catarrh of the stomach, drowsiness, watery secretions, thirst, dryness of the skin.

Dose: Five celloids every two hours.

SUGGESTIONS

The natural functions of elimination should be kept active by taking when necessary a mild laxative remedy. The diet should be light but nourishing. Fresh air and exercise suitable to the patient's condition is very desirable.

If there is an obstruction of the bile ducts, this should not be allowed to remain long, as it has an injurious effect upon the liver. The services of a competent physician are needed in such conditions.

KIDNEY DISORDERS

CAUSES AND SYMPTOMS

Probably the most prevalent kidney disorder is nephritis (also called Bright's disease). It is an inflammatory condition of certain tissues of the kidneys, which if it progresses causes destruction of the kidney cells, impairing the proper function of this vital organ.

In the beginning the trouble is marked by a chill, rise of temperature, decrease in the amount of urine excreted and a change in its composition, and if uremia develops the disease may end fatally.

The chronic form of nephritis has less severe symptoms, the progress of the trouble is slower, but recovery occurs less frequently than from acute nephritis.

Either form of nephritis is too dangerous a disease to justify a treatment without the services of a reliable physician.

LARYNGITIS

CAUSES AND SYMPTOMS

Laryngitis is an inflammation of the vocal organs and the membrane of the upper part of the windpipe. The disorder frequently begins as a common cold with gradual extension of the inflammation from the nose to the throat. The characteristic symptoms are hoarseness, dryness of the membranes, roughness of the vocal cords. In some cases the trouble becomes chronic with chronic cough. Deficiencies in *Ferrum Phos.* and *Kali Mur.* are important factors in this condition.

TREATMENT WITH THE SCHUESSLER REMEDIES

Ferrum Phos.—Painful hoarseness due to strain on the vocal cords or from taking cold; soreness of the larynx, fever.

Kali Mur.—Loss of voice, croupy cough with expectoration of thick, white, tenacious mucus. Tongue white.

Kali Phos.—Hoaresness after nervous strain; tired, weary feeling, general weakness.

Calcarea Phos.—Chronic hoarseness with much hemming and scraping of the throat. Aggravation from talking.

Dose: Five celloids every hour during acute condition, at longer intervals as improvement occurs.

SUGGESTIONS

In acute laryngitis the vocal cords should be given a rest. A cold compress placed over the throat sometimes relieves acute congestion, as will hot drinks.

Smoking is prohibited. If the condition does not speedly improve, a medical examination should be made in order to detect the possible presence of foreign bodies or growths on the larynx.

CLINICAL REPORT

Severe case of acute laryngitis, voice husky and hoarse, cough irritating and painful, dryness of membranes, dark red. The tonsils were swollen, red; temperature 102½°. I prescribed *Ferrum Phos.* to be taken every hour. In twenty-four hours the fever was gone, and the other symptoms were relieved considerably. After two days the tonsils were normal in appearance, the cough loose and painless. On the fourth day recovery was complete and the man returned to his business.

(D. D. C., M.D.)

LEUCORRHEA

CAUSES AND SYMPTOMS

A catarrhal condition of the mucous membranes of the female organs with abnormal discharges, which vary from clear water to thick, offensive secretions. They may be bland or excoriating (scalding) to an extent as to destroy the upper layer of the skin. Aside from mineral salt deficiencies contributing causes are: infections, anemia, general weakness and run-down conditions, relaxation of the tissue of the female organs with resulting displacements.

Persistent discharges, especially when blood-tinged, demand a careful medical examination in order to discover the possible presence of a malignant condition.

TREATMENT WITH THE SCHUESSLER REMEDIES

Kali Mur.—Leucorrhea, when the discharge is thick, milky-white, mild, non-irritating mucus.

Kali Sulph.—Discharge of slimy or watery yellow mucus, sometimes tinged with green.

Natrum Mur.—When the discharge is watery, scalding, irritating or smarting, with itching of the parts or morning headache. Patient is gloomy, tearful, depressed.

Calcarea Phos.—As a constitutional tonic. Discharge thick, clear, transparent, looks like white of egg before it is cooked, albuminous; worse after menses, with weakness in sexual organs.

Natrum Phos.—Creamy, golden-yellow, or acid and watery, acrid or sour-smelling discharges.

Kali Phos.—Scalding and acrid discharges which are traceable to a nervous condition.

Silicea.—Profuse discharges; leucorrhea instead of the menses, in weakly, poorly nourished, or scrofulous constitutions.

Dose: Five celloids every two hours.

SUGGESTIONS

Cleanliness of the parts is always demanded and the use of mild antiseptic, but non-irritating douches are desirable. A mild solution of some Creozone Antiseptic together with about 20 celloids of the indicated remedy in a quart of water should be used as a douche. The water for a douche should be about body temperature.

If there is anemia the treatment suggested for this condition should be adopted.

LUMBAGO

CAUSES AND SYMPTOMS

Lumbago is a painful ailment affecting the muscles of the small of the back. The pains are usually of stabbing character, aggravated when attempting to move. The initial acute attack is sometimes followed by a soreness of the muscles. A deficiency in certain mineral salts, which are required for the maintenance of normal metabolism of the muscles is the basic cause and exposure to cold, over-strain and toxemia from infections are contributing causes.

TREATMENT WITH THE SCHUESSLER REMEDIES

Ferrum Phos.—Early stage, fever, inflammation, severe pain.

Natrum Phos.—Lumbago with acid condition; sour, acid perspiration, rheumatic tendency.

Calcarea Phos.—Backache and stiffness from slightest draft, worse in the morning.

Calcarea Fluor.—Lumbago following a strain.

Dose: Five celloids every one-half to one hour during acute attacks, less frequently as relief is obtained.

SUGGESTIONS

Absolute rest in bed is desirable in acute attacks of lumbago. Pain usually can be relieved by hot, dry applications. Plasters and analgesic balms also act soothingly. Light massage with balm, when the muscles are not too sensitive, will relieve the stiffness. When the patient is able to get out of bed, suitable plaster applied across the sore muscles will relieve pain and allow walking without much discomfort. Provide for free elimination using mild laxatives if necessary, and drinking plenty of pure water to promote elimination through the kidneys.

MALARIA

(Intermittent Fever)

CAUSES AND SYMPTOMS

Malaria is caused by an organism which penetrates into the blood corpuscles, and, multiplying there, finally destroys the corpuscles. This organism is introduced into the body through the bites of mosquitoes which have previously bitten patients affected with malaria. This disease, therefore, is preventable and can be stamped out by destroying the breeding places of this species of mosquitoes and by screening malarial patients from their bites.

Malaria is characterized by periodic attacks of chills, fever and sweat in sequence occurring daily, but more often only every second or third day.

A predisposing cause of malaria is cell-salt deficiency, and the destruction of the blood-cells causes further deficiencies of the salts needed to rebuild the cells. This replacement must take place before the patient can be restored to normal health. Quinine is used effectively for the purpose of destroying the organism in the blood stream, but quinine does not correct mineral salt deficiencies, and the Schuessler Remedies can be used in addition to quinine without in any way conflicting with the effectiveness of quinine used for the destruction of the malarial germ.

TREATMENT WITH THE SCHUESSLER REMEDIES

Natrum Sulph.—The principal remedy in all stages; especially indicated for the bilious symptoms which may arise, bilious stools, greenish coating of the tongue, bitter vomiting.

Ferrum Phos.—An important remedy for the fever, especially when there is vomiting of undigested food. Should be taken in alternation with *Natrum Sulph.*

Natrum Mur.—Fever, severe throbbing headaches, intense thirst, aching all over body.

Kali Mur.—In alternation with *Natrum Sulph.* when the tongue has a thick white coating.

Natrum Phos.—Acid conditions, sour smelling perspiration, vomiting of sour masses.

Dose: Five celloids every two hours.

SUGGESTIONS

During the chill the patient should be wrapped in blankets and given hot drinks. When the fever comes on give sponge baths and cooling drinks. Following the sweat give alcohol rubs and provide dry linen. The diet should be light but nutritious. During and after convalescence a suitable Tissue Tonic should be given for a considerable period until complete strength is regained. Severe or protracted cases of malaria sometimes leave the patient in an anemic condition. (See treatment suggested for anemia.)

CLINICAL REPORTS

Mr. P. W., age 21. Had a severe chill at 11 A. M. He had had chills for three days before he came to me. I put him on *Natrum Mur.*, four tablets every three hours, and *Natrum Sulph.*, five tablets three times a

day. The next chill was very light and there were no others.
(DR. H. W.)

Mr. R., age 47. Was attacked with malaria, having worked in malarial tracts. Had chills and fever in the first attack. Was given *Natrum Sulph.*, three celloids every hour with a little hot water during fever and three times a day at other times. Had no chill in the second or subsequent attacks. The fever period also gradually lessened and the patient was cured in six weeks. Diet—non-stimulant, vegetarian food without any fatty or fried viands.
(DR. P. G.)

Boy X, age 14 years. Had malarial fever for two weeks, temperature ranged up to 104° F. When I saw him he was enormously swollen, eyes and mouth watery, tongue slightly whitish in middle. *Natrum Mur.* every three hours. Much better in four days. *Calcarea Phos.* acted well as a restorative.
(R. M. T., M.D.)

MARASMUS

SYMPTOMS

Marasmus is a disease of infancy and occurs generally only in artificially fed infants.

The child seems unable to properly assimilate food, generally due to mineral deficiencies which disturb the process of assimilation. The child loses weight, gradually wastes away, the muscles and the abdomen are weak and flabby. The child eats but gets no benefit from the food.

TREATMENT WITH THE SCHUESSLER REMEDIES

Calcarea Phos.—Is the principal remedy in this disease. Child emaciated, abdomen sunken and flabby, dentition delayed. Colic after eating, diarrhea.

Natrum Phos.—If there are acid symptoms present, some vomiting. Alternate with *Calcarea Phos.*

Kali Phos.—Wasting of the muscles. Putrid-smelling stools. When there are nerve symptoms present such as prostration, listlessness, sleeplessness, nervousness, etc.

Natrum Mur.—When there is rapid emaciation of the neck, in children, accompanied with irritability, constipation; alternate with *Calcarea Phos.*

Dose: Three celloids five times daily. Dissolve the celloids in the milk or food or in a teaspoonful of water. *Calcarea Phos.* is required in nearly all cases and should be given to the child every two or three hours in alternation with other indicated Schuessler remedies.

SUGGESTIONS

Proper feeding, that is, a well-balanced diet including fruit juices and thick vegetable soups, is required. White of egg in warm water alternated with milk or other foods will provide suitable nourishment. Salt baths are also beneficial.

CLINICAL REPORTS

Mrs. G. called me to her home to see her eight-months-old baby. The child was very much emaciated, neck so small and weak that it could not hold its head up. The abdomen was sunken, and the legs were not much larger than a broom handle. The child had vomiting and constipation, was blue about the eyes and face, had prominent veins, and the skin was dry and wrinkled, more marked about the face and legs.

This child had been fed on condensed milk. This was changed to "Lactogen" and baby given *Calcarea Phos.* before each feeding until improved, and then three times a day. After still further improvement, once a day, and after that at intervals as necessity required. The baby made a complete recovery.

(J. D. V., M.D.)

Some time ago I treated a marasmic baby not a year old with *Calcarea Phos.* and had remarkable success. The mother was profuse in thankful gratitude, for she had tried many remedies and patent foods without results.

For delayed eruption of teeth, loss of weight, and a scrawny condition it stands at the head of the list.

(H. T. D., M.D.)

A boy, F. G., one year and ten months old. Inherited predisposition to tuberculosis by maternal origin. Delayed development, first steps at 16 months. The boy was sick four months before without any relief, and his mother brought him to me in the following state:

Sixteen pounds in weight; 29 inches tall; only five incisors, three above and two below; clean and red tongue; big head covered with sweat; sinking eyes; cyanotic, blue circles around eyes and mouth; wax-white pallor, dirty skin, and at the level of the fontanels depression by lack of consolidation; emaciation; pulse weak; percentage of hemoglobin forty (Fleischel hemometer).

Frequent stools, eight to ten in 24 hours with the following characteristics: green, mucus with particles of undigested food, white mushy and very offensive, with little spots of suppuration; offensive flatus, aggravated after eating and in every sudden change of atmosphere. He was found better when his mother allowed him to rest.

In such conditions I take the characteristic features of the remedy as follows: lack of consolidation of the fontanels, delayed teeth, clammy and cold sweat in the face and skull; green, mushy, and very offensive stools, with undigested food, flatulence worse by ingest food.

I gave him *Calcarea Phos.*, three tablets every three hours, and it relieved the boy in three weeks. He is now in quite good health, his weight has increased, and he has twelve teeth.

(Dr. H. L. C.)

MEASLES

CAUSES AND SYMPTOMS

Measles is an acute contagious disease quite dangerous when the patient is under one year old. Fortunately this disease does not occur frequently in infants.

The period of incubation runs eight to twelve days, that is, it takes that long from the time of contact or exposure for the first symptoms of the disease to appear. In the beginning the disease symptoms are like a cold in the head with watery discharge from the nose and eyes, redness of the mouth and throat, tickling cough and the appearance of a few red spots on the palate and inside of the cheeks. After three or four days an eruption appears first on the face and head, consisting of isolated red spots, pinhead size and raised slightly above the surface. This eruption spreads rapidly covering the entire body and limbs in from 24 to 36 hours. The fever, usually present at the beginning, falls with the appearance of the rash, which fades in about five days, after which the outer layer of the skin over the red spots falls off as a fine scale.

Measles is contagious during the entire course of the disease, but more so in the early stages, before the eruption appears. More dangerous than the disease are the possible complications, such as bronchitis, pneumonia, kidney and ear troubles. Whenever there are indications of complications of this nature a physician should be promptly summoned.

TREATMENT WITH THE SCHUESSLER REMEDIES

Ferrum Phos.—For the inflammatory symptoms, fever, congestion, redness of eyes, flushed face. First remedy, and should be continued as long as there is fever.

Kali Mur.—Second stage, swelling of the glands in the throat or neck, cough, white or grayish-white coating on the tongue. Hardness of hearing; loose, light-colored stools, etc. Alternate with *Ferrum Phos.* when there is fever.

Kali Sulph.—If the rash should be suppressed too soon, this remedy will assist in re-establishing it. Skin harsh and dry; to promote perspiration.

Natrum Mur.—Excessive flow of tears or other watery conditions, or when the tongue is coated with frothy bubbles of saliva, thirst, dry skin, much itching.

Calcarea Phos.—After measles as a tonic.

Dose: Five celloids every hour during the first stage; after the eruption has appeared, every two hours.

SUGGESTIONS

The bedroom (isolated if possible) should be well ventilated, with shades drawn to subdue the light, as the eyes are sensitive. The diet should be liquid during the fever, and light for some days until the convalescence is well established. The patient should have plenty of drinking water if it does not cause vomiting. Bathe the eyes frequently with a solution of boric acid. If the itching is intense, rubbing the skin lightly with cocoa-butter will give relief. Keep the bowels open During and after convalescence from measles give a good tonic.

MENINGITIS

CAUSES AND SYMPTOMS

An inflammation of the membranes of the brain; usually results from infection and frequently is a direct extension of an abscess, especially in the ear.

The disease usually begins with chill, fever, headache, stiffness of the muscles of the neck and back. As the inflammation progresses, convulsions and delirium set in. Vomiting is persistent and the fever becomes very high.

Meningitis is too serious a disease to justify home treatment without medical supervision and, therefore, a physician should be consulted without fail.

Until arrival of the doctor cold packs or ice bags can be applied to the head, and the patient should be kept absolutely quiet in a darkened room. Frequent doses of five *Ferrum Phos.* celloids may be given in the early stages, for the fever.

MENOPAUSE

(Change of Life)

The period of a woman's life during which the menses cease constitutes a critical time of her existence, and one in which great care must be taken that damage is not done, particularly to her mind and nervous system.

The onset of the change of life is usually gradual, beginning around the forty-fifth year and continuing from one to three years, and is marked by irregularity in the appearance of menstruation, gradual diminution in the flow, with more or less nervous disturbances, such as mental unrest, irritability.

TREATMENT WITH THE SCHUESSLER REMEDIES

Ferrum Phos.—Poor circulation of the blood, hot flashes, and chills, sensation of heat on top of head. Congestion during period of menses. Vomiting of undigested food, headaches.

Kali Phos.—Nervousness, mental depression, irritability, anxiety, fainting spells.

Natrum Phos.—Indigestion, acid or sour symptoms.

Calcarea Phos.—General weakness, run-down condition, anemia, loss of weight.

Dose: Five celloids four times daily.

SUGGESTIONS

The patient must have nourishing food, plenty of fresh air and sunshine and moderate exercise. A change of scene and surroundings is often advantageous if the mental disturbances predominate. Mental comfort is important, hence special consideration in this respect should be given to the patient by members of the family and friends. The bowels should be kept regular and, if the patient in anemic suitable Tissue Tonic should be given regularly.

MENORRHAGIA

(Profuse Menstruation)

CAUSES

When the flow of blood at the menstrual period is too profuse and of too long duration the condition is called Menorrhagia. This may be brought about by increased congestion of the pelvic organs, by too fluid blood (non-coagulating) due to a lack of certain mineral salts or by a laxity of the muscles of the uterus due to a *Calcarea Fluor.* deficiency. Anemia is a contributory cause in some cases.

TREATMENT WITH THE SCHUESSLER REMEDIES

Ferrum Phos.—Menses too frequent and too profuse, painful, blood bright red, congestion.

Kali Mur.—Black, dark, clotted blood, catarrhal conditions, leucorrhea.

Calcarea Fluor.—Menorrhagia with bearing-down pains, also when associated with displacement of the uterus (falling of the womb).

Calcarea Phos.—Menses too early in young girls, also when anemia is present.

Dose: Five celloids of the indicated remedy every two hours.

SUGGESTIONS

When the flow of blood is excessive the patient should remain quiet in bed.

When menorrhagia occurs frequently, rest before time for the menses is helpful. If anemia should be present it should be corrected (see article in this book on Anemia).

If menorrhagia persists it is advisable to submit to a careful pelvic examination, especially if the patient approaches the menopause (change of life).

CLINICAL REPORT

Had a patient suffering from profuse menorrhagia. Her family doctor had treated her for three months, but the condition remained unchanged.

Put her on *Ferrum Phos.* Within three days her bleeding stopped and she was able to sit up in a chair. In another day she was walking around the house.

Once since then she has had another attack, but the same treatment gave prompt relief.

(Dr. A. H.)

MENTAL DISTURBANCES

CAUSES

A variety of disturbances of the mentality may occur from different causes. Some of the causes are associated with other diseases, others may occur spontaneously in the absence of disease elsewhere in the body.

Business, financial and family worries are potent factors in the production of mental disorders; so are mental overwork or too great fixation of the mind on one subject. Physical weakness also has been known to lead to mental weakness. Any condition which causes a loss of the important nerve cell-salt, *Kali Phos.*, is likely to weaken the activity of the brain and cause mental disturbance. It might produce only a minor disturbance such as fatigue of the mind or a more serious condition such as insanity.

A competent physician should be consulted whenever possible for the treatment of mental disorders. The correction of mineral salt deficiencies, however, is always desirable and is capable of producing beneficial results, and justifies their use in cases where home treatment must be relied upon.

INDICATIONS FOR THE USE OF THE SCHUESSLER REMEDIES

Kali Phos.—The principal Schuessler Remedy in mental disturbances, nervousness, irritability, depression, loss of memory, sleeplessness, despondency, etc.

Ferrum Phos.—Congestion in head, rush of blood to the head, dizziness.

Natrum Sulph.—Mental disturbance associated with biliousness.

Calcarea Phos.—General weakness. Mind wanders, incapacity for concentrated thought, dullness.

Natrum Mur.—Sadness, depression, readiness to weep, anxiety about health.

Dose: Five *Kali Phos.* celloids every two to three hours. When other remedies are also indicated, take them in alternation with *Kali Phos.*

SUGGESTIONS

Mental rest and change of environment will bring an improvement in the patient's condition. Associated ills must be corrected before definite improvement of the mental condition can be expected. If the patient, for instance, is anemic and undernourished he should be built up to more normal condition. Home surroundings in some cases are not conducive to improvement and the best that can be done is to provide institutional care. (See treatment suggested for anemia.)

CLINICAL REPORTS

Miss E., age 29 years. Office work. Nervous, tired out, easily excited, depressed, fleeting pains over body. Pale. More or less menstrual irregularities. Work an effort. Headaches come and go.

Kali Phos., five celloids every two to four hours. Relieves quickly and seldom has to be taken more than a few days. (ROY C. W., M.D.)

Miss S., aged 18 years, high school scholar, began worrying about her final examinations two weeks ago. Cannot concentrate her mind on her studies, wakeful nights, talks in her broken sleep about her lessons, would get up in the morning so tired she could "hardly drag herself to school." A few doses of *Kali Phos.* celloids acted like a charm, sleeps well, concentrates on her studies, has no fear of the examinations, and wants to know what was in those little tablets to give her so much "pep." (J. B. K., M.D.)

One special case I had while in Steubens Sanitarium. A mental case. The parents were thinking they would be obliged to put the patient in a sanitarium. Dr. Walker, head physician of the establishment, consented to my trying the Schuessler Remedies. I gave her *Kali Phos.* She soon began to show improvement and finally became normal.

(DR. H. F. H.)

Young girl of nervous mental type. Anxious to head her class in school. Was nervous and sleepless from over-study. A few doses of *Kali Phos.*, occasionally, kept her in school, and she graduated at the head of her class.

(J. M. S. C., M.D.)

MUMPS

CAUSES AND SYMPTOMS

Mumps is a contagious disease, believed to develop from bacteria which find suitable breeding ground in tissues deficient in the vital mineral salts. Lack of *Ferrum Phos.* allows inflammatory symptoms to develop and deficiency of *Kali Mur.* produces excess fibrin, a medium for the bacterial growth and a cause of congestion and swelling in the parotid glands.

The incubation period, that is, the time elapsing from the time of infection until the appearance of the first symptoms of mumps, is from 14 to 21 days. The beginning (after the incubation period) is marked by chill, fever, vomiting and general ill feeling. Twenty-four hours later the parotid gland (located behind the angle of the jaw) becomes swollen and painful, and usually soon thereafter the opposite gland also becomes involved. Acid foods or drink (oranges, lemons, etc.) cause a marked aggravation. The swelling subsides in about a week or ten days. In adults the disease is occasionally severe, sometimes shifting from the parotid gland to the breasts or ovaries in the female and to the testicles in the male. This condition may become extremely painful and requires enforced rest until well.

TREATMENT WITH THE SCHUESSLER REMEDIES

Ferrum Phos.—In the early stage; fever, pain, flushed face, vomiting.

Kali Mur.—Second stage, marked swelling of the gland, tender to touch, tongue covered with thick white coating.

Natrum Mur.—Excessive flow of saliva. Painful swelling of the testicles. (Alternate with *Ferrum Phos.*)

Dose: Five celloids every one to two hours during the early stage, at longer intervals thereafter.

SUGGESTIONS

Heat applied over the swollen glands in the throat usually brings relief. This can be accomplished very satisfactorily by applying some suitable Ointment, as hot as can be borne, cover with plenty of cotton, and hold it in place with a bandage. Care should be taken to prevent chilling, which in adults may influence transfer of the inflammation to the ovaries or testicles. If the swelling does shift to these regions, a physician should be called in.

NEURALGIA

CAUSES AND SYMPTOMS

Neuralgia is the medical term applied to pain along the course of a nerve, caused by pressure, irritation or secondary to mineral salt deficiencies. It is generally a sharp, stabbing or burning pain, coming in paroxysms. The tissues about the affected nerve are sore and sensitive to touch.

A severe and persistent form of neuralgia involves the facial nerves of one side of the face. These attacks occur periodically. Severe cases of neuralgia are sometimes designated as neuritis. This is not correct, however, as neuritis is a more serious inflammatory condition of nerves, which if not controlled progresses to destruction of the nerve tissues, with resulting paralysis of the muscles associated with the affected nerve.

TREATMENT WITH THE SCHUESSLER REMEDIES

Magnesia Phos.—The chief remedy for all neuralgic pains; intense, darting, excruciating or spasmodic pains; pains such as are relieved by heat and aggravated by cold.

Kali Phos.—Neuralgic pains in nervous, sensitive persons. Pains are better after gentle motion. Pains with depression, failure of strength, nervousness, sleeplessness, irritability, crossness, etc. Feeling of numbness.

Natrum Mur.—Severe neuralgic pains, intermittent and with excessive flow of saliva or tears. The pains resemble those of *Magnesia Phos.*, but are distinguished from them by the excessive secretions of fluids from the mucous membrane of some organ.

Ferrum Phos.—Neuralgic pains due to inflammatory conditions, caused by a chill or cold; severe throbbing pains, like a nail being driven in over the eye; blinding pains, with fever, burning heat, flushed face.

Calcarea Phos.—Pains coming on at night and of a numbing character, or with sensation of crawling or coldness, deep-seated pains, neuralgia in anemic persons.

Dose: Five celloids of *Magnesia Phos.* every 15 to 30 minutes during acute attacks, after relief is obtained five celloids of the indicated remedy every two hours. If more than one remedy is required, use them in alternation.

SUGGESTIONS

Most neuralgias are aggravated by cold and relieved by heat. Hot water bottles, hot salt bags or an electric light held near the painful parts may bring relief. Analgesic balm applied to the painful area is a desirable palliative. Some cases stubbornly resist medicinal treatment and sometimes require surgical attention.

CLINICAL REPORTS

Mrs. P., age 80 years, fat and of a rather sallow complexion. For the past 30 years has suffered with rheumatism, especially in hands, knees and feet. Had bunions on both feet. Feet were flat and calloused. Heart enlarged. Has suffered with tic-douloureux for 20 years. Had only gotten relief by taking substantial doses of codein.

Present condition was about as above, some pain nearly every day, but at times very bad, so that she held on to her face and rubbed it to get relief. Much worse eating, got some relief by cold applications. After trying what seemed like indicated remedies without results, I decided to put her on a general nerve tonic. Gave her *Kali Phos.* After about a week's treatment she claimed to feel improvement, so continued with the remedy for a long time. Within six months her pain had entirely ceased and she lived to be 86 years old without a return of the complaint.

As a general nerve tonic, given with general symptoms, I have never found a remedy to give better results.

(R. L. E., M.D.)

I have used the Biochemic Cell-Salts in a number of cases with good results. But the one that stands out in my mind was a cure with *Magnesia Phos.* A woman about 47 years of age, heavy build, mother of three children, suffered for a number of years with a severe pain in the right arm, sometimes unable to move arm, and unable to sleep on account of the soreness. Had tried all kinds of treatments and medicines without results. Was taking electric treatments and Infra-Red ray applications when I happened to be in the neighborhood where she lived and met her husband whom I had known for years and had not seen for quite some time. He seemed very much worried about his wife's suffering and asked me if I knew of anything that I thought would give relief. I gave him a bottle of *Magnesia Phos.* and told him to give his wife five celloids every half hour in hot water until bed time and continue as long as she was awake. He did this, and, to my great surprise, he called me up the next morning and told me that the pain and soreness left some time during the night.

(Dr. F. X. C.)

I found this patient with the curtains drawn, the room darkened, patient lying helpless in bed unable to raise up in bed or to be raised up by his nurse. On being raised up in bed he would swoon with pain.

I ordered five *Magnesia Phos.* celloids every 30 minutes until some relief of pain, and then to widen the intervals between the doses. I then went home, leaving the patient seven miles out in the country.

Upon calling at the house the next morning I found the curtains raised, the room flooded with sunshine, and the patient able to raise up in bed without any headache whatsoever. No pain on pressure over the frontal sinuses, temperature normal, pulse normal. Patient desired food after third dose of medicine and had taken only a few doses through the night. And no medicine except *Magnesia Phos.* had been given since the evening before when it was begun. Before *Magnesia Phos.* was started aspirin combined with codein, morphine and various coal tars, etc., had all been given without results. (Dr. I. M. H.)

A case of post-influenzal neuralgia (supraorbital). After the patient seemed to be getting fairly over the fever, and, as I thought, would not need another visit, I was called to him in a hurry to relieve a pain over the eye. I knew the value of *Magnesia Phos.* in neuralgia, but at the time I was out of it, having just sent an order to Luyties Pharmacal Company for a supply. I gave him morphine hypodermically, but with no relief. The man suffered terribly. He was running subnormal temperature, the room had to be darkened, so dark that one had to grope one's way to where he was. On the least admission of light he would be wild with pain.

Toward night of the third day his father announced his intention to call a surgeon as he thought this was a true sinusitis. I asked him to wait for one more mail, because I had some medicine coming in which I had great faith. I received the medicine at 8 P. M.

As soon as I could reach his home I gave him five celloids of *Magnesia Phos.* every 30 minutes until he was easier, then five every hour until he was fully at ease. They told me next day that after the fourth dose of the medicine he was resting and sleeping so well that they would not awaken him to give more medicine, and that he slept soundly and awoke next morning free from pain.

I kept him on the *Magnesia Phos.*, however, for two or three days longer, and he has never felt the pain since.

(Dr. I. M. H.)

Magnesia Phos. is doing fine work for me in treatment of agonizing pains that accompany muscular spasms, especially of the involuntary muscles. Some physician said within my hearing last year that when you felt morphine to be an urgent necessity, try *Magnesia Phos.* first—with hot water at frequent intervals, to insure prompt absorption. It has been especially valuable to me in the treatment of intestinal and uterine colics. I value it especially as a prophylactic against the tendency to such muscular spasms.

(Dr. A. L. M.)

NEURASTHENIA

CAUSE AND SYMPTOMS

Neurasthenia is the medical term applied to a run-down condition of the nervous system. This run-down condition may be slight and be characterized by irritability and nervousness, or it may be more severe and result in a complete breakdown of the nervous system. Mineral salt deficiencies, especially of Potassium Phosphate (*Kali Phos.*) and various contributory causes such as prolonged physical or mental strain, worries, etc., are the chief causes of neurasthenia.

In some cases the functions of the organs become disturbed, especially those of the stomach, heart, kidneys and brain. The trouble is gradual in onset and recovery therefrom is correspondingly slow and difficult, and supplies a real problem to the most competent physician.

TREATMENT WITH THE SCHUESSLER REMEDIES

Kali Phos.—Depressed spirits, irritability, loss of memory, headaches, sleeplessness. *Kali Phos.* is an important remedy and tonic in most cases of neurasthenia.

Calcarea Phos.—Neurasthenia in anemic and constitutionally weak persons.

Ferrum Phos.—Congestion to the head, flushed face.

Magnesia Phos.—Neurasthenia with acute nerve pains, painful contraction of muscles.

Natrum Sulph.—Nervous debility, with bilious condition, dyspepsia.

Dose: Five celloids every two hours. *Kali Phos.* should be used in alternation with other indicated remedy or remedies.

SUGGESTIONS

Various general measures, suited to every individual case, are of great importance in the treatment of neurasthenia. In severe cases the judgment of an experienced physician in such matters is almost indispensable. As a general rule rest is the most important measure; mental rest is required more than physical rest, unless the patient is weak, run down or undernourished. In such cases particularly, a suitable Tissue Tonic should be given : two teaspoonfuls after meals and at bedtime. Provide diversion of mental and physical activity and if the condition of home surroundings worries and irritages the patient, he should if possible be placed in new surroundings.

The diet should be very substantial and the food be invitingly prepared to stimulate the appetite and desire for food. Bowels should be kept open.

A cold spinal douche in the morning is considered helpful.

CLINICAL REPORTS

Woman, 40 years old, suffering from extreme nervousness during menopause. Could not sleep, and was cross and irritable. Nothing pleased her and was in constant family jars. *Kali Phos.* relieved and everybody was happy.

(Dr. A. H.)

A lawyer, J. V., 45 years old, married, seven normal children, good antecedents. Wassermann negative, normal urine, slight arterial hypertension; stature, five feet nine inches; weight 170 pounds; has tobacco habit, non-alcoholic, sedentary life; hard, intellectual work.

He came to my clinic with severe troubles of nervous origin, itching and intense pain in lower extremities, sharp neuralgias at level of the deltoid muscles, biceps and pectorals spasmodic muscular contraction with neuralgia, shooting pains along the nerves, radiating pains in the forearms of both arms. Showed, also, agitation, anxiety, insomnia from exhaustion.

My guides for the administration of the medicine are the leading symptoms: forgetful, incapacity to think clearly, change of one to another idea, increase of pains after the least mental work, excruciating, terrible neuralgic pains shooting like lightning, changing place along the muscles and nerves, extraordinarily tired, languid, and all the pains are relieved temporarily only by warm applications.

I gave him *Magnesia Phos.*, six celloids every three hours. After the first dose there was marked relief. Two weeks more of the medicine and the symptoms disappeared so he could again start his professional work.

(Dr. H. L. C.)

Girl, age 15. Party very nervous. Unable to attend school. Would cry out in sleep. Gas when eating many foods. Constipated. Unable to keep quiet.

Laboratory test. Urine, spec. gr. 1039; acidity 50 per cent; urea 2.2.

Party placed on *Natrum Phos.,* five celloids every three hours, *Kali Phos.* five celloids for the nerves, and one dose per day of *Calcarea Phos.,* four celloids. In one month's time party well and all right after one year.

(DR. F. H.)

Mr. M. Aged 55 years, 5 feet 8 inches in height, weight 170 pounds. Occupation foreman over a department in piano factory. Enjoyed good health until about two years ago when he began to notice that when any little thing went wrong in his work, or if anyone did not treat him "just so" it worried him. He lost self-confidence, would go to bed and could not sleep on account of worrying. Began to have cramps in his fingers and toes, then in arms and legs, would have to get up and take a hot bath for relief. Lost appetite, and, of course, some weight.

Urine normal, slight constipation, diagnosis—neurasthenia.

Mr. M. consulted me on May 1st. I gave him a 4-dram vial of *Kali Phos.* and a 4-dram vial of *Magnesia Phos.,* four celloids every two hours in alternation. He reported in five days, sleeping much better, cramps almost gone. Continued same treatment another five days, no more worrying, no more cramps, has confidence enough in himself now to run the whole factory.

(J. B. K., M.D.)

Miss G. F., age 27. Employed as a typesetter in job printing office. Neurasthenia: Became despondent, wanted to be alone, broke into tears while talking, lost all interest in life, had lost several pounds, and could not eat or sleep. I prescribed outdoor walks with pleasant company and *Kali Phos.* five celloids every two hours. After three weeks she experienced a marvellous change, was cheerful, could eat and sleep well, and was gaining flesh. She returned to her work and has had no further trouble in past two months. Still taking *Kali Phos.* every few hours. I could enumerate a number of cases so treated with brilliant results. I am satisfied it will pay any physician to investigate and use the Biochemic Remedies. They are very pleasant and effective. (Dr. J. W. K.)

NEURITIS

CAUSE AND SYMPTOMS

As explained in the article on Neuralgia, the disease known as Neuritis is characterized by an inflammation of the nerves, progressing to destruction of the nerve structure. The pains in neuritis are usually constant, unlike the spasmodic spells of pain in neuralgia. A single nerve may be involved, especially when caused by injury. When many nerves are affected, the condition is called multiple neuritis, which in the majority of cases has a toxic background (poisonous substances in the system).

TREATMENT WITH THE SCHUESSLER REMEDIES

Ferrum Phos.—Early stage, congestive and inflammatory condition, severe pain, fever. Neuritis from exposure to cold.

Kali Phos.—Lancing pains, worse at rest and beginning to move, shrinking of the tissues, paralyzed sensation in the parts.

Magnesia Phos.—Severe, acute pains, failing strength in muscles.

Dose: Five celloids every one to two hours, according to the severity of the pains.

SUGGESTIONS

Absolute rest is required. In severe neuritis rest in bed. The diet should be simple, non-stimulating, but nourishing. Plenty of pure water assists the elimination of toxic materials through the kidneys. Hot applications are desirable and give relief, but quicker results are obtainable with an analgesic balm. Proper massage of paralyzed muscles with this balm is useful to prevent shrinking. If the patient is in a rundown physical condition, Luyties Tissue Tonic should be given.

PAIN

CAUSES

Pain is not a disease in itself, but is a symptom present in the majority of disease conditions affecting the human body. Every tissue in the body is supplied with nerves, and any disturbance in the tissues is transmitted to the brain as a painful sensation.

Pain is therefore primarily a warning signal that a disease process is going on somewhere in the body, and while uncomfortable, is frequently a valuable sign. Pain is one of the best guides to the progress of the disease, and it is not always desirable to deaden it by sedatives, as this destroys the evidence of the true condition of the patient.

The correction of mineral salt deficiency, as a rule, does not only palliate the pain, but assists in the correction of the underlying cause of pain. It is therefore the natural and rational way to relieve pain. Very severe pain, due to injury, etc., should of course receive the attention of a physician.

TREATMENT WITH THE SCHUESSLER REMEDIES

Ferrum Phos.—Throbbing pain in congestions and inflammations with flushing of the skin at the painful area. In the early stage of painful acute disorders, especially when there is fever and rapid pulse.

Magnesia Phos.—Sharp, stabbing, shooting pains, especially along the course of the nerves. All neuralgic pains. The type of pain which is relieved by heat and aggravated by cold.

Kali Phos.—Pain causing lameness or loss of strength in the affected parts. In persons who are nervous, excitable.

Kali Mur.—Pain associated with swelling of the affected parts.

Calcarea Phos.—Numbness and coldness of the affected parts, creeping sensation. The patient being in anemic or in a weakened, generally run-down condition.

Dose: Five celloids every 15 minutes to one hour according to the severity of the pain. At longer intervals as relief is obtained. Muscular, nerve and rheumatic pains can frequently be relieved with local analgesic applications.

PARALYSIS

CAUSES AND SYMPTOMS

Paralysis is the loss of the power of voluntary motion, either complete or partial. It may occur as the result of brain hemorrhage (apoplexy), from an injury, or follow certain infections.

An exhaustion or degeneration of the nervous system may follow a deficiency of *Kali Phos.* or *Magnesia Phos.*, the mineral salts needed for proper functioning of nerve tissues. A deficiency of these salts remedies the nerves incapable of transmitting impulses along the tract. Sudden onset of paralysis is usually due to apoplexy.

The prospects for recovery from certain forms of paralysis are not good. On the other hand, with proper treatment under supervision of a competent physician, material improvement, or even complete recovery, may be obtained in cases of acute palsies and infantile paralysis.

The correction of mineral deficiencies is desirable and beneficial in all abnormal conditions, and the use of the Schuessler Remedies does not as a rule conflict with the general treatment, but in serious diseases such as paralysis, the treatment should take place under the personal supervision of a physician.

PLEURISY

CAUSES AND SYMPTOMS

Pleurisy is an inflammation of the membrane which lines the chest wall and covers the lungs. It directly results from germ infection of the tissues deficient in mineral salts, which lowers resistance and provides a medium suitable for the nourishment and growth of the organism.

Motion caused by breathing or coughing produces sharp, tearing or cutting pain, first limited to one side of the chest and frequently extending to the other side later on. There is fever present. In many cases the pleural cavity fills up with serum (fluid) which has to be withdrawn. This, of course, requires the services of a physician. In some cases very little liquid accumulates in the pleural cavity and is gradually absorbed and requires no tapping.

The liquid should not be permitted to remain too long in the cavity on account of the danger of infection by pus germs which cause suppuration and a dangerous relapse. This disease should be treated under personal supervision of a physician.

EMERGENCY TREATMENT WITH THE SCHUESSLER REMEDIES

Ferrum Phos.—For the first stage, fever, pain, shivering; rapid, full pulse; short cough; oppressed breathing; stitch in the side.

Kali Mur.—Second stage, when accumulation of fluid in the pleural cavity has taken place; white-coated tongue.

Calcarea Sulph.—Third stage, with formation of pus in the cavity, slimy yellow coating of the tongue.

Calcarea Phos.—Intercurrently through the course of the disease.

Dose: Five celloids every hour during the early stage, every two hours as the symptoms become less acute.

SUGGESTIONS

Complete rest in bed. In some cases it will become necessary to place the patient in a sitting position to ease the pain caused by breathing. Some suitable ointment should be applied to the chest, as hot as possible, and covered with plenty of cotton held in place by a bandage. The diet should be light but nourishing, low in table salt to limit effusion. If effusion of serum to the cavity develops rapidly, with shortness of breath, be sure to call in a physician at once.

PNEUMONIA

CAUSE AND SYMPTOMS

Pneumonia or inflammation of the lungs is caused by infection of a virulent germ. It is an exceedingly dangerous disease, with a high mortality rate. Home treatment should never be relied upon, and this disease is described here so that it may be recognized and if necessary some emergency treatment may be given until a physician can take charge of the patient.

The onset of pneumonia is usually sudden, beginning with a sharp chill, followed by a rise of the temperature (fever) running to 104 or 105 degrees, headache, pain in the chest, cough with expectoration of mucus which is rusty colored (about the third or fourth day). The breathing is short and rapid, the breathing rises to double or triple the normal rate, causing much strain upon the heart. The so-called crisis occurs from the sixth to the ninth day.

EMERGENCY TREATMENT

Ferrum Phos.—In the early stage, five celloids every one-half to one hour.

Kali Mur.—When there is white, thick expectoration.

Kali Sulph.—Dry skin—to promote perspiration.

Keep the patient warm in bed, in a well ventilated room, give light, liquid food only. Principally hot drinks. The bowels should be kept open, preferably by enemas. Protect the chest from drafts and exposure with a warm cotton jacket, changing as often as it becomes moist with perspiration. Some suitable ointment applied hot to the chest, and covered by the cotton jacket, several times daily, is beneficial.

PREGNANCY

During pregnancy a woman is required to carry on the processes of life in her own body and in addition to provide all the material and life building processes for the formation and development of a new life. This places an added burden on her and necessitates an increased supply of the mineral salts, which are often not sufficiently abundant in the diet, but can be supplied independent of the food. Added to this strain is the stress of modern life with its influence upon the nervous system, with the result that what should be a natural function of life becomes often a critical period of a woman's life, with anxiety and a variety of annoying disturbances, culminating in a unnaturally difficult labor.

A careful analysis of the existence of possible mineral deficiencies and the correction thereof before the disturbance has caused too great damage will bring the expectant mother to the termination of her pregnancy in good mental and physical condition to undergo labor without complications and unnecessary hardship.

This procedure is also helpful in preventing the much-dreaded child-bed fever.

An abundance of the vital mineral salts, especially of *Ferrum Phos.* and *Calcarea Phos.* is a great aid to the proper development of the child prior to its birth.

The minerals have also a definite field in the management of disorders occurring during pregnancy and labor.

INDICATIONS FOR THE SCHUESSLER REMEDIES

Kali Phos.—During the entire period of pregnancy, especially if there is evidence of nervous strain.

Ferrum Phos.—Morning sickness during pregnancy, vomiting of undigested food. After-pains severe and long lasting. Also in so-called milk fever, flushed face, restlessness.

Calcarea Phos.—A valuable remedy during pregnancy, well to take during the entire course of pregnancy to aid normal development of the child. Weakness, weariness of the mother during and after pregnancy, poor assimilation of food.

Natrum Mur.—Morning sickness with frothy, watery phlegm.

Natrum Phos.—Morning sickness with vomiting of sour mucus, acidity of the stomach.

Natrum Sulph.—Vomiting in pregnancy, of bilious matter, bitter taste.

Dose: Five celloids every one-half to one hour in acute conditions, less frequently after amelioration. If more than one remedy is required use them in alternation.

SUGGESTIONS

The diet during pregnancy should be nutritious, easily digestible, but not stimulating. Moderate exer-
bowels when sluggish. Avoid compression of the breasts with tight clothing, develop the nipples by gentle massage and absolute cleanliness; prevent cracked nipples. Whenever possible engage a competent physician or obstetrician.

CLINICAL REPORTS

For prenatal treatment, unless some other remedy is evidently indicated, it is my custom for several years to prescribe *Kali Phos.,* a dose twice a day for three months or more previous to confinement. It seems to improve the general health, allay nervousness, and promote normal excretion. Also, the same remedy when labor begins, a dose every one-half hour or hour. Labor is facilitated and shortened. The pains are mitigated.

(Dr. L. C. S.)

It is not my desire to call your attention to the pathological conditions in this critical yet most interesting period of a woman's life. Childbirth should be practically painless, but, thanks to our so-called high state of civilization, it is far—very far—from the ideal. Suffering is the rule, and it has come to be looked upon as the proper procedure, in accordance with the biblical threat, "In sorrow thou shalt bring forth children."

The child of nature—the Indian squaw—suffers but little in labor, although she may have sinned more than her white sister. But the white sister, though she be ever so pure in heart and clean in body, has, through her adherence to the exacting demands of so-called civilization, violated so many laws of nature. With fear of the agony to come, she looks forward to the hour of delivery with anything but pleasurable anticipations, and, as the first pangs of labor begin, she inquires anxiously for chloroform, "Twilight Sleep," to annul her sufferings. This is altogether wrong, but it is a part of civilization. We cannot, however, turn back the hour glass of thousands of years and place woman on the pedestal from which she fell; the conventional sins of ages have left their scars, but happily Biochemistry offers much to mitigate the result of violated law.

It holds out a hope of which millions have never dreamed, the hope of relief from the acute agonies of child-bearing.

My experience and that of other physicians is that *Kali Phos.*, taken for a month or two prior to the expected time of delivery, will add much to the comfort of the expectant mother. It will build and strengthen the nervous and muscular system to such an extent that delivery will be accomplished, in many cases, with comparative ease and comfort.

(Dr. J. B. C.)

PROSTATIC DISORDERS

CAUSES AND SYMPTOMS

Prostatic disorders, especially enlargement of the prostate gland, are conditions which occur most frequently in elderly men. The causative factors are somewhat obscure, but there is little doubt that mineral salt deficiencies play an important part. The most annoying feature of prostatic enlargement is the difficulty it causes in voiding the bladder. Generally a progressing aggravation necessitates the use of the catheter to relieve the pressure on the bladder. In cases where the obstruction becomes very severe an operation may become advisable.

Medicinal treatment is essentially palliative, but the use of the Schuessler Remedies may also retard the progress of the disorder.

TREATMENT WITH THE SCHUESSLER REMEDIES

Ferrum Phos.—Acute prostatic congestion, irritation, retention of urine or difficult urination. Fever.

Calcarea Fluor.—Chronic enlargement of the prostate gland. Testicles hard and enlarged, muscles relaxed, flabby.

Calcarea Phos.—Chronic condition, dribbling of prostatic fluid. General run-down condition.

Dose: Five celloids every two hours.

SUGGESTIONS

Careful and simple living is required. Prevent indigestion and constipation and colds. (See articles on these subjects.) Drinking of considerable quantities of pure water is desirable, but avoid alcohol.

CLINICAL REPORT

I am not in general practice now. I have one case that may be of interest to you. Adult male about 60 years of age, general health satisfactory, complains of a cold sensation or feeling in perineum; diagnosed by two other physicians as enlarged prostate, dense or hard to touch (rectally). No difficulty in urinating. Feeling of depression mentally, sadness.

I ordered three celloids of *Luyties Calcarea Fluor.* four times a day. Result—much improvement. Generally feeling better, sensation of warmth at perineum, —this after three weeks' treatment.

(Dr. A. P. F.)

RHEUMATISM

CAUSES AND SYMPTOMS

Rheumatism is a term applied to a variety of conditions where there is pain, swelling, or inflammation of muscles or joints. Acute rheumatism or so-called inflammatory rheumatism comes on suddenly and is one of the most painful forms, so severe as to make it difficult for the patient to endure the weight of light bed clothes. Gout is a condition affecting chiefly the great toe.

Chronic rheumatism sometimes occurs after repeated attacks of acute rheumatism. It causes a permanent enlargement of the joints, a condition which is usually very difficult to correct.

As in most diseases, mineral salt deficiencies are also an important factor in rheumatic conditions. Even some of the exciting causes of rheumatism such as infection from local pus accumulations, improper functioning of the stomach and intestines, excess of uric acid in the blood, etc., can be traced to mineral salt disturbances. The correction of rheumatism is frequently dependent upon removal of some obscure underlying condition, a problem which may tax the resources of the best physician. The treatment for chronic rheumatism must be followed persistently and for long periods.

TREATMENT WITH THE SCHUESSLER REMEDIES

Ferrum Phos.—In the first stages of rheumatism, for the pain, fever, heat, redness, quickened pulse—rheumatic fever. Rheumatism located in any part; pains are increased by movement; soreness and stiffness all over the body. (Use in frequent doses.)

Kali Mur.—Second stage of rheumatism or rheumatic fever, when swelling has resulted (alternate with *Ferrum Phos.*). Tongue coated with thick, white or grayish coating. Swelling of the joints. Movement increases the pains. Also chronic rheumatism, with swelling of the parts.

Natrum Phos.—One of the principal remedies in certain types of rheumatism (acute or chronic), with sour-smelling perspiration or acid conditions; acid taste in the mouth; creamy-yellow coating on the base of the tongue.

Kali Sulph.—Useful when the pains shift suddenly from one place to another—"wandering" rheumatism. Pains are worse in a heated room or in the evening; feel easier in cool air.

Kali Phos.—Acute or chronic rheumatism, with stiffness of the parts or associated with nervous conditions;

pains worse on beginning to move, but relieved by continual gentle motion.

Natrum Mur.—Chronic rheumatism of the joints with characteristic watery discharges. Rheumatism worse at night, in bad weather, with heat or cold or change of weather. Rheumatism with cold, numb sensations.

Natrum Sulph.—Chief remedy in chronic gout. In the acute stage alternate with *Ferrum Phos.* Rheumatic pains associated with bilious symptoms, and which are worse in wet weather.

Calcarea Fluor.—Enlargement of the joints from rheumatism.

Magnesia Phos.—Acute, sharp spasmodic pains in rheumatism; excruciating, violent pains relieved by warmth.

Dose: Five celloids every hour during acute attacks, less frequently as relief is obtained. In chronic conditions five celloids four or five times daily. When more than one remedy is required, take them in alternation.

SUGGESTIONS

Any harmless non-injurious measure which can give the sufferer relief from pain is desirable. Massage with an analgesic balm, covered afterwards with a warm cotton bandage, or the application of a suitable plaster to the painful area, are useful for this purpose in most cases. As rheumatism is genereally aggravated by cold and dampness it is important that the patient be well protected by warm, dry clothing, whether up and dressed or confined to bed.

The diet in acute rheumatism especially if there is fever, should be light and non-stimulating. In chronic rheumatism the diet should be liberal, nourishing and well balanced. Red meats should be eaten in only moderate quantities. Bowel elimination should be properly regulated, as an inactive colon is frequently a source of infection.

Massage with an analgesic balm is of value in chronic cases as it tends to increase the local circulation and prevent the solid deposits in the joint tissues. Hot baths also promote circulation and therefore are beneficial.

CLINICAL REPORTS

B. S. Age 41 years. Crossing watchman. Out all kinds of weather. Neurasthenic type. Several attacks of flu and pneumonia. Severe headaches—aggravated by change of weather. Rarely any trouble in summer. Came to me November, 1930, with rheumatic pains in left shoulder joint. At time could hardly move arm. Pain sharp—then ache. Kept awake at night. Nervous, restless and cross. Gave him *Ferrum Phos.* and *Magnesia Phos.*, five celloids of each three times per day. Saw him once last winter for same trouble and none since. (R. C. W., M.D.)

Lady, age 80, suffered many years with pain in legs, arms and back with more or less severe swollen, burning feet and legs, also varicose veins. She took *Natrum Phos.*, *Natrum Sulph.* and *Natrum Mur.*, five celloids alternately every two hours for a few months.

(Dr. E. B. G.)

Lady, age 62. Hand and arm greatly swollen and painful. She took *Ferrum Phos.* and *Natrum Phos.*, five celloids alternately every two hours; in two days' time the pain and swelling were not so severe, and after continued treatment normal condition was the result.

(DR. E. B. G.)

RICKETS *(RACHITIS)*

CAUSES AND SYMPTOMS

Rachitis (Rickets) is a disease chiefly affecting the bones, and is mainly confined to artificially fed infants, and due to a lack of vitamines and mineral salts, especially Phosphate of Lime (*Calcarea Phos.*).

The early symptoms are tenderness of the muscles causing the child to cry when picked up, soft, flabby muscles, then faulty development of the bones, slow teething, enlarged head, breast-bone projects, limbs emaciated, and various other deformities.

After deformities have developed it is difficult or impossible to correct these conditions by means of medicinal and dietetic treatment, and surgical treatment becomes necessary.

Rachitis, however, is preventable, and if early attention and proper treatment is given, good results can be obtained. Obviously the best results are attainable by a treatment under the personal supervision of a physician.

TREATMENT WITH THE SCHUESSLER REMEDIES

Calcarea Phos.—Is the chief remedy in this disease, owing to the disturbance in the lime molecules. It is indicated in rachitis with the characteristic symptoms mentioned above.

Natrum Phos.—For non-assimilation of food, with acid symptoms. It also assists in the deposit of the phosphate of lime. (Alternate with *Calcarea Phos.*).

Silicea.—When there is profuse sweat of the head or offensive diarrhea. In alternation or intercurrently with *Calcarea Phos.*, the principal remedy.

Kali Phos.—For the extreme debility often accompanying rickets; also for emaciation, putrid-smelling diarrhea, etc. (Alternate with *Calcarea Phos.*).

Dose: Three celloids dissolved in a little water or in the milk at least five times daily.

SUGGESTIONS

A rachitic child should be kept in the open air and sunshine as much as possible, and the diet should provide all the essential elements. The steady use of *Calcarea Phos.* and preparations containing Vitamin D are important in the treatment of this disease. Cold sponge baths and exercise are also beneficial.

CLINICAL REPORT

Baby Sam, having arrived into the world via the Caesarian route, thirteen months prior to the time he was brought to me by his little mother who weighed 90 pounds and of poor health and unable to nurse little Sam, was fed on goats' milk. Little Sam was truly a pitiful specimen of humanity, weighing eleven pounds, of very fair complexion, large head with open fontanels, scabby scalp, brittle hair, emaciated neck and large flabby abdomen, face pale and hollow with sweating about the head when sleeping. Swelling of the submaxilary glands, no teeth in sight at this age. Stools white and sour, sometimes resembling chalk. Swelling of the cervical glands causing pain and a puny cry when turning the neck. Weakness of the limbs. Unhealthy, ulcerated skin.

These conditions resulting from malnutrition with rachitic affections in general cleared up in a few months after the administration of the Schuessler Remedies, *Calcarea Phos.* and *Silicea,* with proper diet. Baby Sam is past two years of age now and is a normal child in every respect.

(J. E. F., M.D.)

SCARLET FEVER

(Scarlatina—Scarlet Rash)

CAUSES AND SYMPTOMS

Scarlet fever is an infectious and contagious disease, generally occurring in children. Adults, however, are not immune unless they have previously had the disease.

The exciting cause is not definitely known, but is believed to be bacteria acting through the respiratory tract where they find a suitable culture medium in the tissues deficient in the vital mineral salts. One attack usually confers immunity.

The period between the implanting of the disease and the appearance of the first symptoms is from five to seven days. It usually begins with fever, sudden vomiting, sore throat followed in 24 hours by a scarlet rash (bright red) spreading rapidly from the neck to the chest, body and limbs. The fever is usually light, the urine scanty and high colored. After a few days the rash fades and the skin peels off in flakes or sheets of skin.

Scarlet fever is contagious from the beginning of the disease to the completion of peeling, and the patient should be completely isolated during all this time. It is one of the dangerous contagious diseases on account of the frequency of complications, especially with lung and kidney troubles.

Whenever possible a physician should be called in and the suggestions for the treatment given here are intended for emergency cases only.

EMERGENCY TREATMENT WITH THE SCHUESSLER REMEDIES

Ferrum Phos.—In the early stage for the fever, quick pulse, headache, shivering, sore throat, bleeding of the nose, and other initiatory symptoms. Alternate with *Kali Mur.*

Kali Mur.—The chief remedy in scarlet fever, to control the disorganized fibrine. Eruption and swelling of the skin, white-coated tongue, albuminous urine.

Natrum Mur.—When there is vomiting of watery fluids, drowsiness and twitching, frothy bubbles of saliva on the edges of the tongue.

Kali Sulph.—To promote the development of the rash and peeling of the skin; to re-establish the eruption when suddenly suppressed. High temperature, dry skin and lack of perspiration.

Calcarea Phos.—Intercurrently during the progress of the disease, and steadily when convalescence has begun.

SUGGESTIONS

The diet must be light, as long as the fever lasts. For the first few days give the child only milk and plenty of water. Keep the mouth and throat clean with a solution of Suitable Creozone Antiseptic. A child who shows symptoms of scarlet fever should be isolated promptly, placed in bed and kept quiet and warm. All clothing worn by the patient should be properly sterilized. In most localities there are regulations that the health authorities must be notified of all scarlet fever cases, and permission must be obtained to release the patient from isolation. When convalescent, the child should be given Somek Tissue Tonic regularly to hasten recovery until complete normal health and strength are re-established.

SCIATICA

Sciatica is neuritis of the sciatic nerve, the large nerve trunk of the thighs and legs. The pain is usually first felt near the hip, but as the nerve becomes more irritated and inflamed the pain extends down to the knee and in severe cases as far as the heel. In most cases only the sciatic nerve of one side is involved.

The causes and the treatment are the same as those for neuritis. (See article on Neuritis.)

CLINICAL REPORTS

Mr. B. has been suffering for seven months with sciatica in left leg, the pain was very severe and fast undermining his health. He has been treated by a very skillful physician all this time, and almost every known remedy was tried, until the physician himself gave up the case and said that he could do nothing more. I was called, found patient suffering with a dull, tensive pain, extending the whole length of the sciatic nerve of the left leg, worse on slightest motion. Prepared a small quantity of *Kali Phos.* in half glass of water, and gave a teaspoonful every ten minutes for an hour, when the pain was much better. Patient then slept until morning. Next night the pain returned; gave same remedy, but with no result. The next night gave *Kali Phos.* and very soon the pain was relieved; continued *Kali Phos.* every two hours, a small dose for a week, and then four times a day for a month. Once during that time had a slight attack, which was soon stopped by putting a small quantity in half a glass of water, and giving a teaspoonful every ten minutes for a while. A year has passed and there has been no return of the trouble. (G. H. M., M.D.)

Man, age 36. When I was called he was in bed. He had a severe pain in left limb from hip to ankle. I put him on *Kali Phos.* every three hours, *Ferrum Phos.* and *Kali Mur.* of each three doses a day, and *Magnesia Phos.* to be taken when the pain was severe. In three days he was out of bed, and on the tenth day he was at work and has had no symptoms of it since.

(Dr. H. W.)

SINUS DISEASE

CAUSES AND SYMPTOMS

The inflammatory condition existing in common colds and nasal catarrh sometimes extends to the channels (sinuses) accessory to the nose. This inflammation causes a swelling of the membranes of the sinuses which produces pain, tenderness over the affected parts, headache, etc. If the channels close up, the secretions may accumulate in the sinuses and cause such distress as to necessitate mechanical procedure to provide drainage. The causes of sinus troubles are fundamentally the same as those of colds and catarrh.

TREATMENT WITH THE SCHUESSLER REMEDIES

Ferrum Phos.—First stage, fever and congestion and pain in the sinus area. Flushed face, rapid pulse, throbbing pain.

Kali Mur.—Dull pain in the sinus area, thick white mucous discharge, stuffiness of head.

Natrum Mur.—Nasal obstruction with watery discharges, loss of sense of smell, inflammation of sinus with sensation of beating as of hammers, worse in cold air.

Kali Sulph.—Yellow, slimy discharges, worse in warm room and in evening.

Silicea.—Chronic condition; thick, offensive, acrid discharges, ulceration of mucous membranes, chronic nasal catarrh.

Dose: Five celloids every hour in acute attacks, less frequently after improvement. In chronic cases five celloids four times a day.

SUGGESTIONS

In order to obtain relief in sinus disease ample drainage from the nose and sinuses is necessary. Local applications capable of shrinking the swollen membranes frequently will serve this purpose.

Attention should be paid to general hygiene. Plenty of exercise in the open, sleeping in well ventilated room (without draft), cold sponging in the morning, and a nourishing diet to build up the patient's vitality are all beneficial measures.

CLINICAL REPORT

Man, age 58, had sinus infection following tooth extraction. When stooping over a most extremely offensive discharge would run from his nose. *Silicea* cured his trouble in a few days. (C. B. G., M.D.)

SKIN DISORDERS

Diseases of the skin are of such a variety that a separate classification and description has not been attempted here. Furthermore, not very much would be gained thereby, for the reason that the name of the disease furnishes no indication for the use of any particular Schuessler remedies; the character of the symptoms alone serves as a guide for the recognition of the deficient mineral salts.

It has been estimated that practically half of all skin diseases are a form of eczema, and a special article on the subject of "Eczema" is given on **page 116.**

TREATMENT WITH THE SCHUESSLER REMEDIES

Ferrum Phos.—In the first stage of skin diseases, when there is fever, inflammation, heat, pain, burning, etc.

Kali Mur.—Second stage of inflammatory conditions. Eruptions on any part of the body or limbs, when the contents are thick and white; generally accompanied with a white-coated tongue, swelling of the affected parts.

Kali Sulph.—All eruptions of the skin, when the discharges are watery, yellow and foul matter. Dry skin; sudden suppression of eruptions; peeling of the skin with or without sticky secretions.

Natrum Mur.—All eruptions of the skin with watery blisters, when the contents or discharges are clear and watery, irritating. The tongue generally is clean, but with bubbles of saliva on the edges. Ill effects from bites and stings of insects.

Natrum Phos.—All skin eruptions, when the discharges are creamy, golden-yellow or colored like honey, irritating; symptoms of acidity and sour smelling perspiration, creamy-yellow coating on the root of the tongue.

Kali Phos.—Eczema and eruptions of the skin, if accompanied with offensive odor; exhausting perspirations; nervousness; secretions cause soreness, itching and crawling sensation, bloody, watery secretions, etc.

Calcarea Phos.—Skin affections, when the secretions consist of albuminous fluid (like the white of an egg before it is cooked). Skin diseases, when associated with anemic conditions. Itching eruptions; pimples on the face at puberty; excessive perspiration, especially about the head (also *Silicea*).

Calcarea Sulph.—Discharge of thick, yellow matter with yellowish scabs; unhealthy wounds with pus with no tendency to heal.

Silicea.—Secretions are mattery, or blood and matter. Offensive odor and sweating of the feet. Perspiration of the head in children. Boils, ulcers with tendency to suppurate. To hasten suppuration.

Natrum Sulph.—Eruptions; the discharges are yellowish and watery; moist skin affections, with yellowish scabs or scales; chafing of the skin. Symptoms are generally associated with bilious conditions.

Calcarea Fluor.—Chaps and cracks of the skin; cracks in palms of hands. Horny skin; suppurations, with hard, callous edges, etc.

Dose: Five celloids every one to two hours in acute conditions. Five celloids four times daily in chronic skin disorders. As in other diseases, two or more of the Schuessler Remedies can be properly used. In such cases it is recommended to take the remedies in alternation.

SUGGESTIONS

In the general treatment of skin diseases, the local use of Calendula Ointment, observation of diet and hygiene are important measures.

In cases where the skin eruption is just a symptom of a disease it is not enough to give treatment for the skin condition only, the underlying trouble must also receive proper attention.

Certain skin troubles, especially poison ivy, eczema, athletes' foot are aggravated if the affected surfaces come in contact with water. In such cases the surface should be anointed with some suitable ointment, which is especially useful for the relief of itching.

Intolerance of certain foods also has a tendency to aggravate some skin diseases. In fact, it is the principal contributory cause of certain skin troubles. Careful observation of the effects of various foods upon the skin condition will enable the patient to gradually eliminate from the diet such foods as cause disturbances or aggravations.

Suitable laxatives should be used tc maintain regularity of the bowel movements and the drinking of liberal quantities of water will promote elimination of toxins through the kidneys.

If serious diseases such as diabetes, tuberculosis or syphilis, etc., are associated with skin eruptions, it is essential to consult a physician.

CLINICAL REPORTS

Mr. G. Age 38. This patient consulted me presenting a marked urticaria of seven days' duration. The characteristic itching and stinging with loss of sleep. The man was on the verge of collapse. I ordered him to bed and prescribed *Natrum Mur.*, five celloids every four hours. After 48 hours I noticed a marked improvement and at the end of the fifth day all lesions and inflammatory condition had disappeared.

(Loy E. R., M.D.)

Mrs. W. Age 49 years. Housewife. Always had fever blisters from colds and from eating corn on the cob. Had her take a few doses of *Natrum Mur.* and trouble vanished very quickly. Immediately relieves the sting and burn and rarely comes to blisters.

(R. C. W., M.D.)

Mr. H. Age about 40 years. Had herpes zoster with internal fever and biting pains. I prescribed *Ferrum Phos.* every hour for first day, then *Kali Mur.* four doses per day for the vesicles which contained a grayish fluid. The complete cure was practically attained in one week. (R. M. T., M.D.)

SLEEP—DISTURBANCES OF

CAUSES AND SYMPTOMS

Sleep is a process of physiological rest of the brain and nervous system. Deviations from the normal sleep are insomnia or inability to sleep and drowsiness or too great desire to sleep.

Most cases of sleeplessness are due to functional disorders of the brain, the normal ebb of blood from the brain does not take place, leaving the brain in a state of activity. On the other hand, exhaustion of the brain causes drowsiness. These disturbances are caused primarily by alterations in the balance of mineral salts in the tissues of the brain and nervous system. Worry, overwork mental or physical, are contributing causes.

TREATMENT WITH THE SCHUESSLER REMEDIES

Kali Phos.—This being the great nerve salt, it is the chief remedy in disturbances of sleep. Sleeplessness following excitement, when due to worry, mental overwork, sorrow, business troubles, excitement or other nervous causes. Yawning and stretching, somnambulism, restlessness, crying and screaming of children during sleep, frequent dreaming.

Ferrum Phos.—Sleeplessness, from feverish congestions. Restlessness, anxious dreams.

Natrum Mur.—Excessive or constant desire to sleep; usual amount of sleep is unrefreshing; patient feels tired and languid in the morning; dribbling of saliva from the mouth during sleep.

Natrum Sulph.—Drowsiness or sleepiness, when the tongue has a grayish or brownish-green coating, bitter taste in the mouth, and when associated with bilious symptoms.

Magnesia Phos.—Sleeplessness, when the brain feels as if it is contracted, arising from nervous causes; spells of yawning.

Dose: Five celloids every one to two hours, less frequently as the condition improves.

SUGGESTIONS

Persons afflicted with insomnia will readily resort to the use of hypnotic drugs. Their prolonged use is almost certain to produce harm, as they act as depressants to the nervous system, besides causing drug habits. If the condition is so severe as to demand the use of such drugs, they should be used only under personal supervision of a doctor.

Physical measures in many cases will prove very helpful. A warm bath before retiring is generally restful, and a glass of hot milk or cocoa may be given. The bedroom should be well ventilated, the bed should have a firm mattress and light but warm covers. In the morning a cool or cold shower or splash bath followed by a brisk rub is also recommended. Coffee, tea and all stimulating food and drink should be eliminated from the diet. If the patient does considerable mental work, some physical exercise should be taken daily, preferably out of doors. Sun-bathing has given pleasing results in some cases. When the patient is run down physically, suitable Tissue Tonic should be taken regularly over a period of several weeks.

SORE THROAT

Simple sore throat, usually preceding a cold, in which there is redness and swelling, together with pain in swallowing, is very frequently met with. It is usually an acute condition caused by exposure to cold, which throws certain of the mineral salts out of balance.

The treatment to be followed is that of the common cold. (See "Colds," page 79) *Ferrum Phos.*, five celloids every half hour, should be given, and the throat gargled frequently with a solution of Luyties Creozone Antiseptic.

Suggestions for the treatment of various other types of "Sore Throat" are given in the chapters on "Laryngitis" and also on "Tonsillitis."

In laryngitis the vocal cords are irritated, there is hoarseness, and a tickling cough, etc. In tonsillitis there is a painful inflammation and swelling on the tonsils, with difficulty in swallowing.

Severe cases with abscess formation are known as quinsy (see article on Tonsillitis).

The description of the symptoms of diphtheria, a dangerous throat trouble, will be found on page 104.

STOMACH DISORDERS

CAUSES AND SYMPTOMS

In most gastric or stomach disorders we have to deal with a disturbed function of the stomach membranes, usually dependent upon mineral salt deficiencies. Changes take place in the make-up of the secretions which are indispensable in the process of digestion, assuming the form of stomach disorders such as indigestion, catarrhal inflammation or gastritis, chronic dyspepsia, etc. A common condition arising from mineral salt deficiencies is hyper-acidity, which if not corrected may develop into stomach ulcers, a quite troublesome disease.

Disturbances of the nervous system frequently react upon the stomach, causing a form of dyspepsia which demands very persistent treatment.

TREATMENT WITH THE SCHUESSLER REMEDIES

Ferrum Phos.—Acute, painful stomach disturbances, especially if there is fever and vomiting of undigested food. Also when there is a burning sensation in the stomach, pain soon after eating. Loss of appetite, aggravation of the disorder from acids, sweets and coffee.

Kali Mur.—All gastric derangements when the tongue has a white or grayish-white coating. Indigestion with pain or a heavy feeling in the region of the liver. Gases and sick feeling, inactivity of the liver, especially after eating fat or rich food, also bitter taste in the mouth; constipation.

Kali Phos.—Stomach disorders associated with nervous disturbance. Indigestion with all gone feeling and nervous depression. Gas gathers under the heart causing distress. Fright or emotional disturbance upsets the stomach.

Natrum Phos.—Gastric disorders with symptoms of acidity, sour risings, acid taste, heartburn, belching of gas with acid taste, loss of appetite; moist creamy yellow coating at the back part of the tongue and palate, especially in the morning. Pain in stomach usually two or three hours after eating.

Natrum Mur.—Stomach troubles with vomiting of clear watery fluid, violent thirst, stringy, frothy saliva; offensive breath.

Natrum Sulph.—Stomach disorders arising from biliousness with bitter taste in the mouth, vomiting of bile or bitter fluid, bilious headache, giddiness. Tongue coated greenish-brown or greenish-gray.

Magnesia Phos.—Acute, griping pains in pit of the stomach, cramps in the stomach, belching gives no relief.

Calcarea Phos.—Pain after eating even small quantities of food, belching relieves temporarily. Dyspepsia with hunger, excessive accumulation of gas. Food is not properly assimilated.

Dose: In acute conditions, five celloids of the indicated remedy should be taken every hour or even more frequently. At longer intervals as the symptoms subside. In chronic gastric disorders, five celloids four times daily. If more than one remedy is required, take them in alternation.

SUGGESTIONS

Dietary errors are the principal contributory causes of gastric disorders and the correction of bad eating habits is a necessary measure in the successful treatment of these troubles.

Not only should the food be thoroughly masticated, but the diet should be balanced, and should not consist of too great a proportion of a single type of food and be deficient in others.

During acute attacks of stomach disorders it is advisable to withhold all food for a while, and to assume a regular diet gradually. Rich, highly seasoned food, coffee, stimulants, etc., are as a rule not suitable foods for patients with stomach troubles. Furthermore, the patient should eliminate from his diet any of the foods which in his experience are causing an aggravation of the trouble.

The use of alkalies (soda bicarb., etc.) for the purpose of neutralizing acidity of the stomach, while it may give temporary relief, is harmful in the long run, and will generally aggravate the diseased condition. Better results as to ultimate recovery are gained by supplying the mineral salts which are found to be deficient, as outlined in this article.

It should be remembered, however, that the road to recovery in old chronic cases of stomach disorders is bound to be slow.

A careful medical examination should be made in case the patient is persistently suffering pains when the stomach is empty.

CLINICAL REPORTS

Woman, 20, is troubled with burning pains in the stomach after eating. Pains come on and off two hours after eating. She was annoyed and dreaded to eat on account of the pains. Bowels loose, tongue a light gray color. Have given *Natrum Phos.* and *Kali Mur.* alternated every two hours. Change at once and continued the treatment and she grew fat and well.

(L. T. K., M.D.)

Miss R., age 30. This lady is a stenographer. Neurasthenic condition from too much work. Everything she ate caused distress, even milk or anything containing milk gave her acute indigestion.

Prescribed *Calcarea Phos.* and *Kali Phos.* alternately, five celloids every hour, and the next day alternating with *Kali Mur.* and *Natrum Mur.* The gastric irritability corrected after one month. Started to gain a little in weight. She has taken the remedies for three months and has been restored to normal and is able to return to a regular diet. Has resumed her work.

(Dr. F. H.)

ST. VITUS DANCE (CHOREA)

CAUSES AND SYMPTOMS

This is a disease of the nervous system causing quick, jerking movement of certain muscles. It is an evidence of a high degree of irritability of the nervous system, and a faulty conduction of nerve impulses, basically depending upon mineral salt deficiencies. Contributing causes are constitutional weakness, overwork, either mental or physical, worry, etc. In mild cases the jerking occurs only during waking hours, but in severe cases the motions continue during the sleep.

TREATMENT WITH THE SCHUESSLER REMEDIES

Magnesia Phos.—This is the chief remedy for the spasms, involuntary movements and contortions of the affected parts of the body.

Calcarea Phos.—Patient in run-down condition, anemic or scrofulous. To be used in alternation with *Magnesia Phos.*

Kali Phos.—For nervousness, restlessness, mental disorders, neurasthenia, etc. Intercurrent with *Magnesia Phos.*

Silicea.—Spasms, distorted eyes, pale face—obstinate cases, jerking during sleep. Alternate *Magnesia Phos.*

Natrum Phos.—If due to worms in children, or if acid symptoms are present.

Dose: Five celloids every two hours.

SUGGESTIONS

Healthful regulation of the patient's living habits, such as simple nourishing diet, exercise in the open air, plenty of sleep, ample mental rest, etc., are helpful in restoring normalcy. Suitable Tissue Tonic should be given regularly for its tonic effects on the system. Other associated disturbances, however, must also be corrected. This cannot always be easily and quickly accomplished, and generally requires the services of an experienced physician.

SUNSTROKE

(Heatstroke)

Sunstroke, more accurately called heatstroke, is a condition of the body produced by exposure to great heat, combined ordinarily with marked humidity of the atmosphere. Sun exposure is not necessary to bring on this trouble, as artificial heat will produce the same result. When the air is hot and moist, perspiration, which ordinarily by its evaporation cools the temperature of the body, does not evaporate, but remains in drops upon the skin, while the body accumulates heat, until its temperature is so high that a heatstroke is brought on.

The symptoms of sunstroke or heatstroke begin with a feeling of oppression and dizziness, followed by sudden unconsciousness, the sufferer lying in deep stupor with heavy, puffing breathing. The face is at first very white and then becomes flushed and red, and the temperature rises to an abnormal height never seen in any other disease. The pupils of the eyes may be contracted or widely dilated. Occasionally, also, this condition will proceed until the sufferer is in a state of collapse, and the head and body will feel cold to the touch. This is known as heat exhaustion.

TREATMENT WITH THE SCHUESSLER REMEDIES

Ferrum Phos.—Five celloids alternated with *Natrum Mur.,* five celloids, every hour and keep the patient quiet.

SUGGESTIONS

The patient should be at once placed in the shade, and all possible clothing removed. To reduce his temperature, if the face is red and flushed, apply ice to the head, and either ice or cold water to the body, rubbing the skin where the water has been applied vigorously meanwhile, which, with the cold water, helps greatly in reducing the temperature. If feasible, the patient may better be placed in a tub of cold water and rubbed, as suggested above.

Complete rest in bed for a number of days after the sunstroke should be insisted on, and the patient should be warned that any exposure to heat for several days will be likely to bring on another attack.

In either sunstroke, heatstroke or heat exhaustion, a physician should, if possible, be called at once.

SYPHILIS

Syphilis is a contagious disease, usually spread by contact with infected people and also by heredity.

The initial sore of syphilis appears a few days after contact in the form of a papule (large pimple) which soon breaks down into an ulcer with hard edges and enlargement of the adjacent glands. This ulcer has a tendency to heal in a few weeks, but is followed shortly by secondary eruptions which may last for weeks or months, then follows a period of apparent improvement, only to be succeeded by a third and frequently fatal sequence of this virulent disease.

The best chance for recovery will be found by the prompt treatment of this disease in its early stages. This treatment should by all means be conducted under constant personal supervision of a competent physician. In order not to give encouragement for self-treatment of this serious disease, no further suggestions on this subject are given in this work.

TONSILLITIS

CAUSE AND SYMPTOMS

An inflammation of the membranes of the throat involving the tonsils. In many cases the inflammation begins in the tonsils and spreads to the adjacent membranes. The tissues weakened by mineral salt deficiencies become non-resistant to the invasion of pathogenic bacteria.

The most prevalent form is acute tonsillitis which begins suddenly with a chill followed by fever, red and swollen tonsils, and general ill-feeling. Swallowing becomes difficult, and there is considerable pain. In the more severe types the tongue becomes heavily coated, the breath is very offensive, and if an abscess forms near the tonsils, the condition is designated as quinsy.

Patients suffering with frequent attacks of tonsillitis should consult a physician as it may be advisable to resort to an operation, which should be performed at a time when the tonsils are not infected.

TREATMENT WITH THE SCHUESSLER REMEDIES

Ferrum Phos.—Early stage, fever, flushed face, rapid pulse, vomiting of undigested food.

Kali Mur.—Second stage, swelling of tonsils, swallowing difficult, tongue white coated. White or gray spots on tonsils.

Calcarea Sulph.—Ulcerating or suppurating tonsillitis, with yellow discharge.

Calcarea Phos.—Chronic enlargement of the tonsils, glands swollen and painful. Pain in throat, especially on swallowing.

Natrum Mur.—Drowsiness, frothy, transparent mucus covering the tonsils, with feeling of obstruction in the throat.

Natrum Phos.—Acid condition of the stomach, creamy yellow mucus on tonsils and base of tongue. Sensation of lump in throat.

Dose: Five celloids every hour in the early stage of acute attacks, then every two hours. For children, three celloids.

SUGGESTIONS

Tonsillitis patients with fever should be kept in bed, on a liquid diet. The throat should be sprayed or gargled fequently with some suitable antiseptic solution. Hot applications of suitable ointment, covered with a cotton bandage, should be made to the throat externally.

If there is an abscess to cause annoyance, it is advisable to consult a physician for the possible lancing of the abscess.

CLINICAL REPORTS

J. W., a groom taking care of horses at fair grounds. History: Many attacks of quinsy. Consulted me about throat and examination revealed tonsils inflamed and badly swollen with engorged veins in pharynx and intense pain and soreness. Ordered *Ferrum Phos.* and *Kali Mur.*, three celloids alternately every two hours. Thirty-six hours later the inflammation was gone and after continued treatment, cure perfect.

(Dr. B. H. B.)

Dr. W. had a severe attack of tonsillitis, involving both tonsils, which were very much enlarged, causing difficult and painful deglutition. Temperature 102; pulse 130; patient exceedingly nervous. Gave *Ferrum Phos.* and *Kali Phos.* in alternation every 15 minutes. The next morning found that the patient had passed a fair night. Continued with *Ferrum Phos.* and *Kali Mur.* In six hours found the patient very much improved; less pain; less swelling; temperature 100; continued the remedies, and in two days the patient was out of bed. The patient remarked that he could feel the effects of the remedies all through his body, quieting and soothing the nervous irritability.

(G. H. M., M.D.)

TYPHOID FEVER

CAUSES AND SYMPTOMS

The exciting cause of typhoid fever is the "bacillus typhosis," usually gaining entrance to the system through the means of infected milk or drinking water, or occasionally infected raw vegetables. The predisposing cause in this as in other infectious diseases is a deficiency in certain of the mineral salts forming a suitable medium upon which the organisms may thrive.

Typhoid fever is too serious a disease to permit self-treatment. A competent physician should be called in promptly for all cases when the symptoms point to the presence of this dangerous illness.

The usual symptoms of typhoid fever are: The first symptoms of this illness appear from two to three weeks after the bacillus has gained entrance into the body. The onset is gradual, there is low fever, headache, general ill-feeling and indigestion. The temperature rises for about four days to a week, remaining high from two to three weeks, and then falling gradually. Weakness, prostration and wasting are in proportion to the degree of fever and inability to assimilate food. Diarrhea is usual, the discharge being watery and often foul and in some cases bloody.

ULCERS

CAUSES AND SYMPTOMS

Ulcers are open sores on the skin surface. The exciting cause is frequently a long continued irritation of the skin or an active injury which does not heal.

If an ulcer does not heal in spite of efforts made along the lines as suggested here, a physician should be consulted as a serious constitutional disease may be the underlying cause, as for instance tuberculosis, syphilis, diabetes, varicose veins, etc.

The use of the Schuessler Remedies in the non-malignant type of ulcers is helpful.

TREATMENT WITH THE SCHUESSLER REMEDIES

Kali Mur.—Ulceration with a thick, white discharge, bland and not irritating. Tongue coated white, base of ulcer is usually swollen.

Silicea.—Base of ulcer is spongy, bleeding readily, hard edges; secretions are thin, yellowish, acrid, pus-like, itching.

Calcarea Fluor.—Deep-seated ulceration. Discharge thin, burning. Ulcers from varicose veins.

Calcarea Phos.—Ulcers in person with scrofulous constitution; in anemic conditions.

Calcarea Sulph.—Chronic ulcers; secretions of yellowish pus with blood. Ulcers following wounds which fail to heal.

Dose: Five celloids every two hours in recent cases, four times daily in chronic conditions.

SUGGESTIONS

Building up of the general health with proper nourishment, exercise, exposure to sunshine, etc., is important in the treatment of ulcers of long standing. Keep the affected parts clean, by bathing them thoroughly with some Creozone antiseptic solution, and cover to protect from injury.

CLINICAL REPORTS

Ethel T., age 10, had bruised her toe on a rock. For three weeks they had tried everything to heal it, but it kept on getting worse. Secretion was of a watery, slightly adhesive nature. Prescribed *Silicea* celloids every two hours. Sore was dried up in 24 hours and completely healed in three days.

(Dr. A. C. N.)

Chronic ulcer of right tibia caused by log falling on leg. After healing being struck by corner of metal-edged box. Suffered for 27 years with no indications of improvement. When first seeing the patient he was 80 years old. The ulcer was almost as large as a silver dollar, with an odor of rotten eggs. The patient had not slept for two weeks as the burning and itching around the edge of the sore was intense. I prescribed *Calcarea Sulph.*, five celloids every three hours. After a few days the healthy granulation began to form. Inside of six weeks the ulcer was completely healed. The man lived to be 89 years old with no more ulcer trouble.

(C. E. S., M.D.)

A young lady, 15 or 16 years of age, came with a slowly discharging sore, posterior to and slightly above the knee, caused by a fall on a small stick a year or two before seeing me. It wouldn't heal, and caused her much trouble. *Silicea* cured the case without further trouble.

(M. A. T., M.D.)

VARICOSE VEINS

CAUSES

The veins are subject to inflammation and to loss of elasticity with resulting weakening and dilation of the walls of vessels, a condition called varicose veins. The principal cause is a lack of *Calcarea Fluor.*, as this mineral salt is necessary to give strength and elasticity to the vein walls. Contributing causes are continuous standing for many hours at a time. It also occurs frequently during pregnancy. Hemorrhoids are a form of varicose veins in the rectum.

TREATMENT WITH THE SCHUESSLER REMEDIES

Calcarea Fluor.—This is the principal remedy when the veins are dilated, also when there is a tendency to varicose ulceration. Bluish discoloration of the tissues.

Ferrum Phos.—Inflammation of the veins, red streaks following the course of a vein, throbbing pain along a vein. Alternate with *Calcarea Fluor.*

Magnesia Phos.—Severe acute pains, cramp-like, spasmodic. Alternate with *Calcarea Fluor.*

Dose: Five celloids four times daily. In acute condition every two hours.

SUGGESTIONS

In severe cases it is necessary to provide artificial support for the weakened vessel. Bandaging the legs with elastic bandages or stockings, usually accomplishes the purpose. Varicose ulcers on the legs can usually be healed by keeping the limbs continuously in an elevated position. After the ulcer is healed the wearing of bandages or elastic stockings will usually prevent the return of the ulceration.

CLINICAL REPORTS

In relaxed condition of the parts like varicose veins, *Calcarea Fluor.* will act magically. I have had several cases within a short time responding very favorably.

(H. T. D., M.D.)

Mrs. H., a lady of 50, came to me with varicose ulcers. Seventeen years before she had an eruption (skin) on her leg. At that time she had varicose veins. Her treatment, of which I do not know the nature, was a failure so far as a cure was concerned. It became an ulcer. She tried various and sundry treatments and methods of treatment, and among them she went to a class hospital. She spent several weeks there, used hot bichloride baths, lay in bed, elevated her limb, and was there several weeks with but little improvement. She went home and drifted along. Her leg grew steadily worse. When she came to me she had great eating ulcers almost from knee to ankle. Her leg was raw in between these sores, almost denuded of cuticle. I put her on *Calcarea Fluor.* and kept her on it. I had her soak her limb in a weak solution of hot alum water twice a day. I had the satisfaction of seeing that leg heal until only one ulcer remained, which was two and one-half inches in length and almost as wide. This sore was greatly reduced in size. At this time I wanted her to put on an elastic stocking, which she positively refused to do. Thereafter I lost track of the patient.

(I. M. H., M.D.)

Case: Woman, Mrs. H., age 54. Light complexion, or of the fat and fair type.

Past history: For several years had suffered with enlarged and painful veins in both legs. Began to have change of life about three years ago, but at times had excessive flowing. Otherwise well.

Present condition: Is flowing quite profusely, and when this ceases has a great deal of hot flashes, so she can hardly go about her work in the house. For over a month has had lot of pain in both legs, below the knees. On examination found all the veins in the locality below the knees distended and purple. On the right leg, at about the bend and at the outer aspect, was a tumor about the size of an egg, which looked as though it might be an infected hematoma about ready to open. I at once gave her *Calcarea Fluor.* and asked her to keep off her feet as much as possible. This she was not able to do as she was taking care of her daughter with a baby. I gave the leg support with bandage. At once she began to improve. That was about last November. Today her veins have shrunk to normal. The hematoma dissolved away without breaking down. The menorrhagia has entirely ceased and the hot flashes have practically disappeared.

(R. L. E., M.D.)

Married lady, age 45, suffering with varicosities of right leg. Swelling, burning pain. Little extravasation of blood into tissues. Slight rise of temperature.

Remedies—*Ferrum Phos.* and *Calcarea Fluor.* alternately three or four days, then continued with *Calcarea Fluor.* three or four times per day. In about six or seven weeks leg was apparently well. Symptoms had disappeared.

(J. S. M., M.D.)

Man, age 55. Scrotal veins greatly enlarged, left testicle sensitive to the touch. General health good. This condition was of twenty years' standing, nevertheless a complete cure was brought about by the administration of *Ferrum Phos.* and *Calcarea Fluor.* celloids within four months.

(R. G. S., M.D.)

VERTIGO

CAUSES

Dizziness with inability to stand or walk, may be temporary and may originate from eye troubles, stomach disturbances or biliousness. Intoxication from drugs or alcohol also cause vertigo. If dizziness becomes persistent, however, a medical examination should be made, as more serious underlying disturbances of the nerves and brain may demand special attention and treatment.

TREATMENT WITH THE SCHUESSLER REMEDIES

Ferrum Phos.—Vertigo or giddiness from rush of blood to the head, with throbbing pain and flushed face. Dizziness, when rising from stooping; after eating or with vomiting of undigested food.

Kali Phos.—Dizziness, when arising from nervous causes, weakness or anemic conditions; worse when rising or looking upward. Alternate with *Ferrum Phos.*

Natrum Sulph.—Vertigo arising from bilious derangements; yellow-coated tongue and bitter taste, with jaundiced condition, denoting an excess of bile in the system.

Natrum Phos.—Dizziness, due to derangements of the stomach, with acid conditions, sour vomiting.

Dose: Five celloids every one to two hours.

SUGGESTIONS

Regularity in living habits should be observed. Correct constipation and other associated disorders. When vertigo is persistent the consultation of a physician is necessary, as early diagnosis and proper treatment are very important.

VOMITING

CAUSES

Vomiting is not a disease but is a frequent symptom in many diseases, especially of those affecting the stomach and intestines. It is an indication of an irritated condition of the stomach, a sign that the stomach is unable to digest and assimilate its contents. It is, therefore, a signal that no attempts for further intake of food should be made until the underlying causes have been corrected. Nervousness is also a frequent cause of vomiting, and it is also a prevalent disturbance during pregnancy.

Persistent vomiting and vomiting of blood may indicate a serious condition, which should be treated under supervision of a physician.

TREATMENT WITH THE SCHUESSLER REMEDIES

Ferrum Phos.—Vomiting of undigested food, sometimes with sour fluids, flushed face, usually fever. Vomiting of bright-red blood.

Kali Mur.—Vomiting of thick, white phlegm. Vomiting, when the tongue is coated white, indigestion worse from rich, fatty food.

Natrum Mur.—Vomiting of sour fluids and watery, transparent fluids. Water-brash. Fluids welling up in the throat, sometimes tasting sour or salty, violent thirst.

Kali Phos.—Vomiting of dark substances like coffee-grounds, bitter, sour taste, nervous condition.

Natrum Phos.—Vomiting of acid fluids or curdy masses. Burning in the stomach (heart-burn). When there is golden-yellow coating on the back part of the tongue.

Natrum Sulph.—Vomiting of bile or bilious matter, with bitter taste in the mouth or other bilious symptoms. Starchy food causes flatulency. Dark coated tongue.

Calcarea Phos.—Vomiting, due to non-assimilation of food; vomiting, recurring regularly at certain hours of the day or night. Vomiting of infants from poor digestion.

Dose: Five celloids every hour in acute attacks, less frequently after improvement. Dose for children, three celloids.

SUGGESTIONS

Withdrawal of all food is usually indicated in vomiting, a measure which may be continued for several days in persistent cases. In some cases placing cold compresses or an ice bag over the region of the stomach will check the vomiting. To wash out the stomach of its irritating contents give large draughts of warm water (one pint) with a pinch of table salt. To clear the lower bowel use warm soapy water for enemas.

WHOOPING COUGH

CAUSE AND SYMPTOMS

Whooping cough is an infectious disease, usually occurring in epidemic form. It is most frequent in children under two years of age. The predisposing factor is deficiency in mineral salts in the tissues of the respiratory tract which results in an accumulation of excess fibrin irritating the bronchial tract and furnishing medium for bacterial growth.

After exposure to the disease a period of about eight to 14 days elapses before appearance of the first symptoms of whooping cough. In the beginning the cough is mild, there are secretions from the nose and throat membranes. The cough gradually increases in intensity, with violent spasms of coughing, with vomiting following attacks. The coughing persists in most cases for several weeks after the end of the period marked by spasmodic attacks.

The greatest danger of whooping cough is the possible complication with pneumonia. If the fever rises in the course of whooping cough and the breathing becomes more difficult between the paroxysms of coughing, a physician should be called to examine the child for signs of pneumonia.

TREATMENT WITH THE SCHUESSLER REMEDIES

Kali Mur.—An important remedy, for the white-coated tongue or thick, white expectoration. Spasmodic cough without the whoop. White, thick phlegm.

Magnesia Phos.—The chief remedy in whooping cough, for the "whoop," paroxysms of coughing, difficult breathing. (Alternate *Kali Mur.*)

Calcarea Phos.—In obstinate cases. Expectorations albuminous, like the white of an egg before it is cooked. In weak, undernourished, emaciated or anemic children.

Natrum Mur.—Expectoration is thin and clear, like water; flow of tears from the eyes.

Ferrum Phos.—Inflammatory conditions, fever; coughing up blood from severe strain. Alternate with the remedy indicated by the expectoration.

Kali Phos.—Intercurrently in whooping cough, with weakness, exhaustion, or in nervous children.

Dose: Three celloids every one to two hours during the active stage, three or four times daily for the cough after spasmodic attacks have ceased.

SUGGESTIONS

Keep the child in the open air in good weather. The sleeping quarters should be well ventilated but free from drafts. The chest should be warmly covered as during the paroxysms of coughing the child will not remain under the bed-clothes, and thus may be chilled. Rub the chest and throat with suitable ointment at night and cover with a cotton jacket or bandage.

Food should be given frequently and in small amounts, as a full stomach increases inclination for vomiting.

A tight belt about the abdomen will often relieve the severity of the spasms of coughing and prevent strain on the muscles of the abdomen, also lessen the tendency to vomiting.

The child must be kept from contact with other children as long as the cough lasts.

CLINICAL REPORT

Child, aged 18 months, in the last stage of whooping cough, with blistered lips and mouth. Black, thin, offensive stools five times a day. Hard and tympanitic abdomen. I prescribed *Kali Sulph.* which effected a complete cure in a short time.

(C. B. K., M.D.)

WORMS

CAUSES AND SYMPTOMS

The intestinal tract is often the site of worms, the eggs of which are taken in with the food.

The presence of worms in the bowel is usually detected by the irritation and itching of the anus. The child becomes nervous, does not sleep well, starts and cries out in sleep, picks at the nose, etc.

TREATMENT WITH THE SCHUESSLER REMEDIES

Natrum Phos.—Is the principal remedy for all kinds of worms; to correct the excess of lactic acid.

Kali Mur.—Small, white thread worms, with itching of the anus, white tongue, etc. (alternate with *Natrum Phos.*).

Ferrum Phos.—Intestinal worms, with passing of undigested food; also for the fever symptoms in all worm troubles. Alternate with *Natrum Phos.*

SUGGESTIONS

The correction of mineral deficiencies will serve to prevent worms, and if they are present it will produce a condition unfavorable to their continued life. For the removal of worms prompt results are, as a rule, obtained with preparations containing Santonin, which deadens the worms (especially pin and round worms) and makes the removal with laxative or salt water enema easy. The enemas should be repeated daily until all signs of the presence of worms have disappeared.

Tape worms are not easily removed (discovered by segments of tape worm), and a physician should be consulted for proper measures.

J. B. Chapman

288 - promote perspiration - kali sulph
247 - dandruff - nat mur, kali sulph
247 - falling out of hair - silica, kali P
loss of hair - calc phos
249 - bitter taste in mouth - nat sulph
251 - depressed - KP, CP, NM
discouraged - nat sulph
252 - poor memory - CP, KP, MP
black spot before eyes - KP
253 - blood shot eyes - FP
256 - humming in ears - NM
257 - ringing in ears - mag phos
258 - itching nose - nat phos
263 - tongue - dry in morning - kali phos

REPERTORY
of the
Application
of the
Twelve
Schuessler Biochemic Remedies

REPERTORY

of the

Application

of the

Twelve Schuessler Biochemic Remedies

HEAD SYMPTOMS

Blind headache: *Ferrum Phos.*

Cold applications relieve: *Ferrum Phos.*

Crawling feeling over head, with cold sensations: *Calc. Phos.*

Crusts, yellow, on scalp: *Calc. Sulph.*

Dandruff: *Natr. Mur., Kali Sulph.*

Dizziness: *Kali Phos.*

Eruptions on scalp, with watery contents: *Natr. Mur.*

Eruption and nodules on the scalp with falling out of the hair: *Silicea.*

" on the head with secretions of decidedly yellow, thin matter: *Kali Sulph.*

Giddiness, with gastric derangements: *Natr. Phos.*

Hair, falling of: *Kali Phos., Silicea.*

" loss of: *Calc. Phos.*

" pulling causes pain: *Ferrum Phos.*

Head, cold to touch: *Calc. Phos.*

" inability to hold up: *Calc. Phos.*

" sore to touch: *Ferrum Phos.*

" sweat on, of children: *Calc. Phos., Silicea.*

" trembling of: *Magnes. Phos., Kali Phos.*

" ulcers on top of: *Calc. Phos.*

Headache accompanied by:

" biliousness, bitter taste: *Natr. Sulph.*
" chills up and down spine: *Magnes. Phos.*
" cold feeling on head: *Calc. Phos.*
" confusion: *Kali Phos.*
" constipation: *Natr. Mur., Kali Mur.*
" dizziness: *Natr. Sulph.*
" drowsiness: *Natr. Mur.*
" dull, heavy hammering: *Natr. Mur., Ferrum Phos.*
" feeling as if skull were too full: *Natr. Phos.*
" frothy coating on tongue: *Natr. Mur.*
" hammering, throbbing: *Ferrum Phos.*
" inability for thought: *Kali Phos.*
" intermittent and spasmodic pains: *Magnes. Phos.*
" irritability: *Kali Phos.*
" loss of strength: *Kali Phos., Calc. Phos.*
" nodules, on head: *Silicea.*
" pain in temples: *Ferrum Phos., Natr. Phos.*
" " over eye: *Ferrum Phos.*
" " in stomach: *Natr. Phos.*
" " throbbing, beating: *Ferrum Phos.*
" " on top of head: *Ferrum Phos., Natr. Sulph.*
" profusion of tears: *Natr. Mur.*
" prostrated feeling: *Kali Phos.*
" rush of blood to head: *Ferrum Phos.*
" sharp, shooting pains: *Magnes. Phos.*
" vomiting of acid sour fluids: *Natr. Phos.*
" undigested food: *Natr. Phos., Ferrum Phos.*

Headache with:

" tearful mood: *Kali Phos.*
" thick white coating on the tongue: *Kali Mur.*
" unrefreshing sleep: *Natr. Mur.*

Headache with:
" vomiting of frothy phlegm: *Natr. Mur.*
" weariness: *Kali Phos.*
" yawning and stretching: *Kali Phos.*

Headache:
" aggravated by mental work: *Calc. Phos., Kali Phos.*
" " in evening: *Kali Sulph.*
" " " heated rooms: *Kali Sulph.*
" from loss of sleep: *Kali Phos.*
" " mental work: *Kali Phos.*
" in nervous subjects: *Kali Phos.*
" neuralgic: *Kali Phos., Magnes. Phos.*
" " with humming in the ears: *Kali Phos., Ferrum Phos.*
" of girls at puberty: *Natr. Mur., Calc. Phos.*
" " nervous character, with illusions of light: *Magnes. Phos.*
" on awakening in the morning: *Natr. Phos.*
" " crown of head: *Natr. Phos.*
" " top of head, with pressure: *Natr. Phos.*
" " " with heat: *Natr. Phos.*
" relieved by cheerful excitement: *Kali Phos.*
" " cool air: *Kali Sulph.*
" rheumatic, evening aggravations: *Kali Sulph.*
" sick, from sluggish action of liver: *Kali Mur.*
" " with bitter taste in mouth: *Natr. Sulph.*

Heaviness of the head in the morning after waking, with giddiness and dullness: *Natr. Mur.*

Inflammatory condition of the scalp: *Ferrum Phos.*

Mouth, bitter taste in: *Natr. Sulph.*

Neck, sharp pain in nape of: *Magnes. Phos.*

Neuralgia of head when pain is sharp: *Magnes. Phos.*

Neuralgic headache with humming in the ears, better under cheerful excitement, worse alone, tearful mood: *Kali Phos.*

Noises in head when falling asleep: *Kali Phos.*

Pain in the nape of the neck of a sharp character: *Magnes. Phos.*

Pain and weight in the back part of the head, with weariness and exhaustion: *Kali Phos.*

Pain, aggravated by cold: *Magnes. Phos.*

" relieved by cheerful excitement: *Kali Phos.*

" " gentle motion: *Kali Phos.*

" " heat: *Magnes. Phos.*

Scalp, eruption on: *Silicea.*

" inflammatory conditions of: *Ferrum Phos.*

" nodules on: *Silicea.*

" painful, pustules on: *Silicea.*

" sensitive: *Silicea.*

" sore to touch: *Silicea, Ferrum Phos.*

" suppurations of, discharge yellow and purulent: *Calc. Sulph.*

" tight sensations of: *Calc. Phos.*

" white scales on: *Natr. Mur., Kali Mur., Kali Sulph.*

Sick headache arising from sluggish action of the liver, want of bile frequently accompanied by constipation: *Kali Mur.*

" when the material vomited is undigested food: *Ferrum Phos.*

" with bitter taste in the mouth; vomiting of bile or bilious diarrhea: *Natr. Sulph.*

" " vomiting of sour fluids: *Natr. Phos.*

Skull, thin and soft: *Calc. Phos.*
Sleeplessness: *Kali Phos.*
Stitches in the head: *Natr. Mur.*
Trembling and involuntary shaking of the head: *Magnes. Phos.*
Vertigo: *Calc. Phos.*
" giddiness from excessive secretions of bile, tongue has a dirty greenish or gray or greenish brown coating at the back part, bitter taste in the mouth: *Natr. Sulph.*
" from exhaustion and weakness: *Kali Phos.*
Violent pains at the base of the brain: *Natr. Sulph.*
Vomiting of white phlegm: *Kali Mur.*

MENTAL SYMPTOMS

Anxious about future: *Calc. Phos.*
Backwardness: *Kali Phos.*
Brain-fag, from overwork: *Kali Phos.*
Children, crossness of: *Kali Phos.*
" crying and screaming: *Kali Phos.*
" ill-tempered: *Kali Phos.*
" peevish and fretful: *Calc. Phos.*
" screaming of, at night, during sleep: *Kali Phos., Natr. Phos.*
" somnambulism in: *Kali Phos.*
Depressed spirits: *Kali Phos., Calc. Phos., Natr. Mur.*
Desires solitude: *Calc. Phos.*
Despondent moods: *Natr. Mur., Natr. Sulph., Silicea.*
Discouraged, feels: *Natr. Sulph.*
Dizziness: *Ferrum Phos., Kali Phos.*
Fainting of nervous sensitive persons: *Kali Phos.*
" tendency to: *Kali Phos.*

Fits of crying: *Kali Phos.*
" " laughing: *Kali Phos.*
Grasping at imaginary objects: *Kali Phos.*
Home-sickness: *Kali Phos.*
Hopeless, with dejected spirits: *Natr. Mur.*
Illusions, mental: *Magnes. Phos., Kali Phos.*
Impatience and nervousness: *Kali Phos.*
Irritable: *Kali Phos.*
Irritation, due to biliousness: *Natr. Sulph.*
Melancholy: *Natr. Mur., Kali Phos.*
Memory, poor: *Calc. Phos., Kali Phos., Magnes. Phos.*
Mind, wanders from one subject to another: *Calc. Phos.*
Moods, anxious: *Kali Phos.*
" gloomy: *Kali Phos.*
Overstrain, from mental employment: *Kali Phos.*
Sensitiveness: *Kali Phos.*
Shyness: *Kali Phos.*
Sleeplessness: *Kali Phos.*
Sleepiness: *Natr. Mur.*
Stupor: *Natr. Mur.*
Thought, cannot concentrate: *Calc. Phos.*
" difficulty of: *Silicea.*
Weeps easily: *Natr. Mur.*

EYE SYMPTOMS

Acrid tears in the eyes: *Natr. Mur.*
Acute pain in eyes: *Ferrum Phos.*
Agglutination at night with smarting of the lids: *Silicea.*
Agglutination of lids in morning: *Natr. Phos.*
Black spots before eyes: *Kali Phos.*
Blisters on the cornea: *Natrum Mur.*
Blurred vision, after straining eye: *Calc. Fluor.*

Burning of edges of lids: *Natr. Sulph.*
Colors before eyes: *Magnes. Phos.*
Contracted pupils: *Magnes. Phos.*
Cornea, blisters on: *Natr. Mur.*
" crusts on eyelids, yellow: *Kali Sulph.*
" inflammation of, with thick yellow discharges: *Calc. Sulph.*
Dimness of sight from weakness of the optic nerve: *Kali Phos.*
Discharge, golden-yellow, creamy: *Natr. Phos.*
" thick white mucus: *Kali Mur.*
" " yellow: *Calc. Sulph.*
" " greenish, serous: *Kali Sulph.*
" " slimy secretions: *Kali Sulph.*
Drooping of lids: *Kali Phos., Magnes. Phos.*
Dry inflammation of eyes: *Ferrum Phos., Natr. Mur.*
Excited appearance of eye: *Kali Phos.*
Eye affections with flow of tears: *Natr. Mur.*
Eyes, blood-shot: *Ferrum Phos.*
" glued together in the morning, with a creamy discharge: *Natr. Phos.*
Eye-balls, ache: *Calc. Phos.*
" pain in the, relieved by resting eyes: *Calc. Fluor.*
Eyelids, specks of matter on: *Kali Mur.*
" yellow, mattery scabs on: *Kali Mur.*
Flow of tears from the eyes when associated with colds in the head: *Natr. Mur.*
" from weakness: *Natr. Mur.*
" on going into open air: *Natr. Mur.*
" with fresh colds: *Natr. Mur.*
" with neuralgic pains in eye: *Natr. Mur., Magnes. Phos.*

Granulations on eyelids: *Ferrum Phos., Kali Mur.*
Inflammation of the eye, acute, with great intolerance of light: *Ferrum Phos.*
Inflammation of the eyes, when pus is discharging: *Calc. Sulph.*
" secreting a golden-yellow, creamy matter: *Natr. Phos.*
" with discharge of thick yellow matter: *Silicea.*
Lids, hot feeling of: *Calc. Phos.*
Light, great intolerance of: *Ferrum Phos., Calc. Phos.*
" sensitive to artificial: *Calc. Phos., Magnes. Phos.*
Neuralgic pains in eyes: *Calc. Phos., Magnes. Phos.*
Neuralgic pain in the eyes with flow of tears: *Natr. Mur.*
Optic nerve, dullness of sight, from weakness of: *Kali Phos., Magnes. Phos.*
Pain as from excoriation in the eyes: *Natr. Mur.*
" in the eyes with tears; recurring daily at certain times: *Natr. Mur.*
Pupils, contracted: *Magnes. Phos.*
" dilated during disease: *Kali Phos.*
Redness and inflammation of the whites of the eyes with sensation as if the eye-balls were too large: *Natr. Mur.*
Sensitive to artificial light: *Calc. Phos.*
Smarting secretions, with tears: *Natr. Mur.*
Sparks before eyes: *Magnes. Phos.*
Sore eyes with specks of matter on the lids or yellow mattery scabs: *Kali Mur.*
Spasms of eyelids: *Calc. Phos., Magnes. Phos.*
Spasmodic twitching of lids: *Magnes. Phos., Calc. Phos.*
Stoppage of tear ducts from cold: *Natr. Mur.*

Squinting: *Calc. Phos., Magnes. Phos.*
" caused by irritation, from worms: *Natr. Phos.*
Staring appearance of eyes: *Kali Phos.*
Sty on lids: *Silicea.*
Weak eyes with tears when going into the cold air or when wind strikes the eyes: *Natr. Mur.*
Yellow crusts on the eyelids: *Kali Sulph.*
Yellow-green matter in the eye: *Kali Sulph.*

EAR SYMPTOMS

Beating in the ears: *Silicea.*
Boils around external ear: *Silicea.*
Catarrh of ear, causing deafness: *Kali Sulph.*
" involving eustachian tubes: *Kali Sulph.*
" " middle ear: *Ferrum Phos., Kali Mur.*
Cracking noises in ear on blowing nose: *Kali Mur.*
" when swallowing: *Kali Mur.*
Cutting pain under ears: *Kali Sulph.*
Difficulty of hearing, accompanied by exhaustion of nervous system: *Kali Phos.*
" accompanied by thick, yellow discharge: *Calc. Sulph.*
" from inflammatory action: *Ferrum Phos.*
" " swelling of eustachian tubes: *Natr. Mur., Kali Mur., Silicea, Kali Sulph.*
Discharges, foul, ichorous, offensive: *Kali Phos.*
" mixed with blood: *Kali Phos.*
" thick, yellow, bloody: *Calc. Sulph.*
Ears, swollen, burning, itching: *Calc. Phos.*
Earache, accompanied by albuminous discharge: *Calc. Phos.*
" beating, throbbing pain: *Ferrum Phos.*

Earache, accompanied by excoriating discharge: *Calc. Phos.*

Earache, gray or white-furred tongue: *Kali Mur.*

" lightning-like pain through ears: *Natr. Sulph., Magnes. Phos.*

Earache, accompanied by swelling of eustachian tube; glands or tonsils: *Kali Mur.*

" yellow, mattery discharge: *Kali Sulph.*

" aggravated by cold: *Magnes. Phos.*

" " damp weather: *Natr. Sulph.*

" of nervous or spasmodic character: *Magnes. Phos.*

" relieved by heat: *Magnes. Phos.*

Exudations from ear, thick, white and moist: *Kali Mur.*

Glands around the ear swollen; noises in the ear; snapping and cracking: *Kali Mur.*

Granulations moist, gray or thick white exudation from the ear: *Kali Mur.*

Heat and burning of the ears with gastric symptoms: *Natr. Phos.*

Humming in the ears: *Natr. Mur.*

Inflammation of the ears, first stage for the fever and pain: *Ferrum Phos.*

" external ear with redness and burning: *Ferrum Phos.*

" loud noise aggravates: *Silicea.*

Noises in ears and head, with confusion: *Kali Phos.*

" like running water: *Ferrum Phos.*

Outer ear sore and scabby: *Natr. Phos.*

" with creamy discharge: *Natr. Phos.*

Scabs, with creamy, yellow appearance: *Natr. Phos.*

Sharp pain under ears: *Kali Sulph.*

Singing or tingling in the ears: *Natr. Mur.*

Stitches in the ears: *Natr. Mur.*

Swelling of the parotid gland with stitching pain: *Silicea.*

Ulceration of the ear when the discharge is foul, ichorous, offensive, sanious, or mixed with blood: *Kali Phos.*

Whizzing and ringing in the ears with diminution of hearing: *Magnes. Phos.*

NOSE SYMPTOMS

Bleeding from the nose: *Ferrum Phos.*

" in delicate constitutions, when the blood is thin, blackish or coagulating; predisposition to bleeding: *Kali Phos.*

" " anemic persons, the blood is thin and watery: *Natr. Mur.*

Boils on edges of nostrils: *Silicea.*

Burning in nose: *Natr. Sulph.*

Catarrh, accompanied by fever: *Ferrum Phos.*

" acute or chronic, with slimy yellow, greenish discharges: *Ferrum Phos., Kali Phos.*

" albuminous discharge, thick and tough, dropping from the posterior nares, causing constant hawking and spitting, worse out of doors: *Calc. Phos.*

" aggravated in evening: *Kali Sulph.*

" " warm room: *Kali Sulph.*

" dry catarrh with stuffy sensation: *Kali Mur.*

" chronic, with purulent discharges from anterior or posterior nares: *Kali Sulph., Silicea.*

" of anemic persons: *Natr. Mur.*

" with fetid discharges: *Kali Phos.*

Catarrh, with salty, watery mucus: *Natr. Mur.*
" " stuffy sensation: *Kali Mur.*
" " white, not transparent phlegm: *Kali Mur.*
Cold in the head with yellow creamy discharge from the nose; itching of the nose: *Natr. Phos.*
" in the third stage of resolution, when the discharge is thick, yellow, purulent, and sometimes tinged with blood: *Calc. Sulph.*
" with dry, harsh skin; to produce perspiration: *Kali Sulph.*
Crusts in the vault of the pharynx: *Kali Mur.*
Discharge, albuminous: *Calc. Phos.*
" clear, watery, transparent mucus: *Natr. Mur.*
" fetid: *Kali Phos.*
" slimy, yellow, watery, greenish: *Kali Sulph.*
" thick and white: *Kali Mur.*
" yellow, fetid: *Silicea.*
" lumpy, green: *Calc. Fluor.*
" purulent, bloody: *Calc. Sulph.*
" yellow, creamy: *Natr. Phos.*
Disposition to take cold in anemic persons: *Calc. Phos.*
Dryness and burning in the nose: *Natr. Sulph.*
Dryness of nose, with scabbing: *Natr. Mur., Silicea.*
Edges of nostrils itch: *Silicea.*
First or inflammatory stage of colds: *Ferrum Phos.*
Fluent coryza: *Natr. Mur.*
Frequent sneezing: *Silicea, Natr. Mur.*
Fresh cold and discharge of clear, watery transparent mucus, and sneezing: *Natr. Mur.*
Hawking and spitting, constant: *Calc. Phos.*
Itching or redness of tip of nose: *Silicea.*
" the nose: *Natr. Phos.*

Loss of smell or perversion of the sense of smell, not connected with a cold: *Magnes. Phos.*
" with dryness and rawness of the pharynx: *Natr. Mur.*
Nose, inflamed at edges of nostrils: *Silicea.*
" swollen: *Calc. Phos.*
Pharynx, dryness and rawness of: *Natr. Mur.*
Picks at nose: *Natr. Phos.*
Sneezing: *Natr. Mur.*
Stuffy cold in head, with yellow, lumpy, green discharges: *Calc. Fluor.*
" with collection of greenish mucus: *Kali Sulph., Silicea.*
Takes cold easily: *Ferrum Phos., Calc. Phos.*

FACE SYMPTOMS

Anemic face: *Calc. Phos.*
Beard, tender pimples under: *Calc. Sulph.*
Blotches on face, come and go suddenly: *Natr. Phos.*
Chaps of lips: *Calc. Fluor.*
Cheek swollen and painful: *Kali Mur.*
Creeping pains in face: *Calc. Phos.*
Dirty-looking face: *Calc. Phos.*
Eruption on the face from any cause, with discharge: *Silicea.*
Face, bloated, without fever: *Natr. Phos.*
" flushed, cold sensation at nape of neck: *Ferrum Phos.*
" livid: *Kali Phos.*
" pale, sickly, sallow: *Kali Phos., Calc. Phos.*
" pallid and pale: *Ferrum Phos., Calc. Phos.*
" red, without fever: *Natr. Phos.*

Faceache, accompanied by:
" constipation: *Natr. Mur.*
" flow of tears: *Natr. Mur.*
" cutting pains: *Magnes. Phos.*
" small lumps on face: *Silicea.*
" from swelling of cheek: *Kali Mur.*
Feeling of coldness or numbness of face: *Calc. Phos.*
Feverish complexion: *Ferrum Phos.*
Frothy bubbles at edge of tongue: *Natr. Mur.*
Grinding pains in face: *Magnes. Phos., Calc. Phos.*
Hard swelling on cheeks, with toothache: *Calc. Fluor.*
Inflammatory neuralgia of the face: *Ferrum. Phos.*
Lightning-like pains in face: *Magnes. Phos.*
Neuralgia, accompanied by flow of tears: *Natr. Mur.*
" " shifting pains: *Magnes. Phos., Kali Sulph.*
" " shooting pains: *Magnes. Phos.*
" " spasmodic pains: *Magnes. Phos.*
" aggravated by being in heated room: *Kali Sulph.*
" " cold: *Magnes. Phos.*
" in the evening: *Kali Sulph.*
" with exhaustion of nervous system: *Kali Phos.*
" relieved by being in cool air: *Kali Sulph.*
" relieved by hot applications: *Magnes. Phos.*
Nodules on face: *Calc. Sulph.*
Pains and heat in face: *Ferrum Phos.*
" cold applications soothe: *Ferrum Phos.*
Pale face in children when teething is difficult: *Calc. Phos.*
" pallid face from a lack of red blood corpuscles: *Ferrum Phos.*

P[illegible]s on face, mattery: *Calc. Sulph.*

" at age of puberty: *Calc. Sulph., Calc. Phos.*

Skin cold and clammy: *Calc. Phos.*

Swellings on face: *Calc. Sulph.*

Tearing pain in face: *Magnes. Phos., Calc. Phos.*

White about mouth and nose: *Natr. Phos.*

Yellow, sallow, or jaundiced face due to biliousness: *Natr. Sulph.*

MOUTH SYMPTOMS

Acid taste in mouth: *Natr. Phos.*

Bad taste in mouth: *Natr. Sulph., Kali Phos.*

" in morning: *Calc. Phos.*

Bitter taste in mouth: *Natr. Sulph.*

Blisters like pimples on the tip of the tongue: *Calc. Phos.*

Canker of lips or mouth: *Kali Mur.*

Clean tongue with an inflammatory condition: *Ferrum Phos.*

Coating on the tongue white and slimy: *Kali Mur.*

" yellow, sometimes with whitish edge: *Kali Sulph.*

Constant hawking of slimy mucus: *Natr. Sulph.*

Constant spitting of frothy mucus: *Natr. Mur.*

Cracked lips: *Calc. Fluor.*

Creamy, golden-yellow exudation from tonsils and pharynx: *Natr. Phos.*

Creamy, yellow coating at back part of roof of mouth: *Natr. Phos.*

Dirty greenish gray or greenish brown coating on the root of the tongue with saliva: *Natr. Mur.*

Dryness of the lower lips; skin pulls off in large flakes: *Kali Phos.*

" " tongue in low fevers with watery discharge from the bowels: *Natr. Mur.*

Glands and gums swollen: *Kali Mur.*

" swelling of, under tongue: *Natr. Mur.*

Gums hot, swollen and inflamed: *Ferrum Phos.*

Hard swelling on jaw-bones: *Calc. Fluor.*

Hawking, constant, of foul, slimy mucus from trachea and stomach: *Natr. Sulph.*

Inflammation of salivary glands, when secreting excessive amount of saliva: *Natr. Mur.*

Mouth full of thick, greenish-white, tenacious slime: *Natr. Sulph.*

Rawness of mouth: *Kali Mur.*

Saliva, excess of: *Natr. Mur.*

Sour taste in mouth: *Natr. Phos.*

Spasms of stammering: *Magnes. Phos.*

Speaks slowly: *Magnes. Phos.*

Swelling of glands under the tongue: *Natr. Mur.*

Thrush in children: *Kali Mur.*

" with much saliva: *Natr. Mur.*

Tongue—see chapter on Tongue.

Twitching, spasmodic, of lips: *Magnes. Phos.*

" mouth: *Magnes. Phos.*

Very offensive breath: *Kali Phos.*

TONGUE AND TASTE

Blisters on tip of tongue: *Natr. Mur., Calc. Phos.*

Chronic swelling of: *Calc. Fluor.*

Clean and red: *Ferrum Phos.*

Coating on tongue, clear, slimy, watery: *Natr. Mur.*
" dirty, greenish-gray, bitter taste: *Natr. Sulph.*
" golden-yellow, on back part: *Natr. Phos.*
" grayish-white: *Kali Mur.*
" like stale brownish liquid mustard, offensive breath: *Kali Phos.*
" moist, creamy, on back part: *Natr. Phos.*
" yellow and slimy: *Kali Sulph.*
Cracked appearance of tongue: *Calc. Fluor.*
Dark red and inflamed: *Ferrum Phos.*
Dry in the morning: *Kali Phos.*
Dryish or slimy: *Kali Mur.*
Frothy bubbles on edges of: *Natr. Mur.*
Induration of tongue, after inflammations: *Silicea, Calc. Fluor.*
Inflammation of, for swelling: *Ferrum Phos., Kali Mur.*
" with exhaustion: *Kali Phos.*
" when suppurating: *Silicea, Calc. Sulph.*
Numbness of tongue: *Calc. Phos.*
Pimples on tip of: *Calc. Phos.*
Swollen: *Kali Mur., Calc. Phos.*
Stiffness of: *Calc. Phos.*
Ulcers on: *Silicea.*
Vesicles on tongue: *Natr. Mur.*

TEETH AND GUMS

Children grind teeth during sleep: *Natr. Phos.*
Cramps during dentition: *Magnes. Phos.*
Decay of teeth as soon as they appear: *Calc. Phos.*
Dentition retarded: *Calc. Phos.*
Enamel, brittle: *Calc. Fluor.*
" rough and thin: *Calc. Fluor.*

Gastric derangements during teething: *Natr. Phos.*
Gums bleed easily: *Kali Phos.*
" pale: *Calc. Phos.*
" predisposition to bleed: *Kali Phos.*
Gum-boil: *Silicea.*
" before pus begins to form: *Kali Mur.*
Infants, teething of, with drooling: *Natr. Mur.*
Nervous chattering of teeth: *Kali Phos.*
Neuralgia of teeth: *Natr. Mur.*
Rapid decay of teeth: *Calc. Fluor., Calc. Phos.*
Seam, bright-red, on gums: *Kali Phos.*
Sockets, teeth loose in: *Calc. Fluor.*
Teeth sensitive to cold air: *Magnes. Phos.*
" " touch: *Magnes. Phos.*
Toothache accompanied by:
" deep-seated pain: *Silicea.*
" excessive flow of saliva or of tears: *Natr. Mur.*
" neuralgia of face: *Magnes. Phos.*
" sharp, shooting pains, spasmodic: *Magnes. Phos.*
" swelling of gums or cheeks: *Kali Mur., Ferrum Phos.*
" ulceration: *Silicea.*
Toothache aggravated by being in warm room: *Kali Sulph.*
" " hot liquids: *Ferrum Phos.*
" in nervous subjects: *Magnes. Phos., Kali Phos.*
" relieved by being in open air: *Kali Sulph.*
" " cold applications: *Ferrum Phos.*
" " " liquids: *Ferrum Phos.*
" " hot applications: *Magnes. Phos.*
Ulceration of roots of teeth: *Calc. Sulph.*
" with swelling gums and cheeks: *Calc. Sulph.*

THROAT SYMPTOMS

Burning sensation in the pharynx and cases of chronic catarrh, when there is considerable dropping from the posterior nares: *Calc. Phos.*

Choking on attempting to swallow: *Magnes. Phos.*

Constricted feeling of throat: *Magnes. Phos.*

Closing of larynx by spasms or cramp: *Magnes. Phos.*

Constant hoarseness: *Calc. Phos.*

Dry, red and inflamed throat: *Ferrum Phos.*

First stage of sore throat, when there is pain, heat and redness: *Ferrum Phos.*

Glands painful, aching: *Calc. Phos.*

Hoarseness, constant: *Calc. Phos., Ferrum Phos.*

Inflammation of the mucous lining of the throat with watery secretions: *Natr. Mur.*

" tonsils: *Ferrum Phos.*

" " with swelling and grayish white patches: *Kali Mur.*

Larynx, burning and soreness in: *Calc. Phos.*

" closing of, by spasm: *Magnes. Phos.*

Loss of voice: *Kali Mur.*

" from strain: *Ferrum Phos.*

Lump in, on swallowing: *Natr. Sulph.*

Pharynx, burning and soreness in: *Calc. Phos.*

Raw feeling in throat: *Natr. Phos.*

Redness and inflammation: *Ferrum Phos.*

Scraping of, when talking: *Calc. Phos.*

Sticking pain in, on swallowing: *Calc. Phos.*

Shrill voice, coming on suddenly while speaking: *Magnes. Phos., Kali Phos.*

Sore, raw feeling in the throat; tonsils and throat inflamed, with creamy, yellow, moist coating: *Natr. Phos.*

" throat as if a plug had lodged in the throat: *Natr. Mur.*

" " of singers and speakers: *Ferrum Phos.*

" " with too much secretion: *Natr. Mur.*

Spasms of the throat: *Magnes. Phos.*

Spasmodic cough: *Magnes. Phos.*

Stinging sore throat, only when swallowing, the neck being painful to touch: *Silicea.*

Suppuration of throat: *Calc. Sulph.*

Swallowing, painful: *Calc. Phos.*

Thirst, with dry mouth: *Calc. Phos.*

Tonsils, chronic enlargement of: *Calc. Phos.*

" creamy, yellow, moist coating on: *Natr. Phos.*

" gray-white patches on: *Kali Mur.*

" inflamed: *Natr. Phos., Ferrum Phos.*

Ulcerations, with thick yellow discharges: *Silicea.*

Ulcerated throat, with fever and pain: *Ferrum Phos.*

" white or gray patches: *Kali Mur.*

Windpipe, spasmodic closing of: *Magnes. Phos.*

GASTRIC (STOMACH) SYMPTOMS

Abnormal appetite, but food causes distress: *Calc. Phos.*

Acid drinks aggravate: *Magnes. Phos.*

All conditions when excess of saliva and watery vomiting present; tongue has a clear, frothy, transparent coating: *Natr. Mur.*

All conditions of the stomach when there are sour acid risings, or the tongue has a moist, creamy yellow coating: *Natr. Phos.*

Appetite not satisfied: *Kali Phos.*

Belching brings back taste of food: *Ferrum Phos.*

" sour eructation: *Natr. Phos.*

Bilious colic: *Natr. Sulph.*

Biliousness from too much bile: *Natr. Sulph.*

Bitter taste in mouth: *Natr. Sulph.*

Bloated, stomach feels: *Calc. Phos.*

Burning in stomach: *Calc. Phos., Kali Mur., Ferrum Phos.*

Catarrh of the stomach with yellow, slimy tongue: *Kali Sulph.*

Clear, frothy, transparent coating on tongue: *Natr. Mur.*

Cold drinks relieve symptoms: *Ferrum Phos.*

" " aggravate symptoms: *Calc. Phos., Magnes. Phos.*

Constipation, with water-brash: *Natr. Mur.*

Craving for salt or salty food: *Natr. Mur.*

Distress about heart: *Kali Phos.*

Dizziness: *Natr. Sulph.*

Dread of hot drinks: *Kali Sulph.*

Dyspepsia with:

" acid risings: *Natr. Phos.*

" flushed face and throbbing pain in the stomach: *Ferrum Phos.*

" pain after eating, if watery symptoms are present: *Natr. Mur.*

Dyspepsia with white gray coating on the tongue, heavy pain under the right shoulder blade, eyes look large and protruding: *Kali Mur.*

Evacuations, bilious, green: *Natr. Sulph.*

Excess of saliva: *Natr. Mur.*

Faint, sick feeling in the region of the stomach: *Calc. Phos.*

Fatty food disagrees: *Kali Mur.*

Flatulence, with distress about heart: *Kali Phos., Natr. Phos.*

Flatulence with sluggishness of the liver: *Kali Mur., Natr. Sulph.*

Food aggravates: *Calc. Phos.*

" causes pain: *Natr. Phos.*

" distresses: *Calc. Phos.*

" persistent vomiting of: *Ferrum Phos.*

Fullness at pit of stomach: *Kali Sulph.*

Great thirst: *Natr. Mur.*

Headache: *Natr. Sulph.*

" with vomiting of food: *Ferrum Phos.*

Heartburn: *Silicea, Ferrum Phos., Calc. Phos., Natr. Phos.*

Heaviness in stomach: *Calc. Phos.*

Hiccough: *Magnes. Phos.*

Hungry feeling after eating: *Kali Phos.*

Indigestion, accompanied by griping pains: *Magnes. Phos.*

Indigestion with pain in the stomach and watery gathering in the mouth, or sour taste in the mouth: *Natr. Mur.*

Indigestion with pressure and fullness at the pit of the stomach: *Kali Sulph.*
" vomiting of greasy, white, opaque mucus: *Kali Mur.*
" watery vomiting and salty taste in the mouth: *Natr. Mur.*
Infants vomit sour curdled milk: *Calc. Phos.*
Liver, cutting pain in region of: *Natr. Sulph.*
Lump, food lies in a: *Calc. Phos.*
Milk, infants vomit curdled: *Calc. Phos., Natr. Phos.*
Moist, creamy, yellow coating on tongue: *Natr. Phos.*
Morning sickness: *Natr. Phos.*
Mouth full of slimy mucous: *Natr. Sulph.*
Nausea, with sour risings: *Natr. Phos.*
Nausea immediately after a meal: *Natr. Mur.*
" with gone sensation in the stomach: *Kali Phos.*
Neuralgia of stomach: *Magnes. Phos.*
Nurse, constant desire of infants to: *Calc. Phos.*
Nurses, child vomits as soon as it: *Calc. Phos., Ferrum Phos., Silicea.*
Pain in stomach after eating: *Natr. Mur., Calc. Phos., Natr. Phos.*
" is remittent and spasmodic: *Magnes. Phos.*
" sometimes relieved by belching: *Calc. Phos.*
" worse from eating even the smallest amount of food: *Calc. Phos.*
Pastry disagrees: *Kali Mur.*
Pressure at pit of stomach: *Kali Sulph.*
Right shoulder-blade, pain under: *Kali Mur.*
Salty taste in mouth: *Natr. Mur.*

Sick headache from gastric derangements: *Natr. Sulph.*
Sour, acid risings: *Natr. Phos.*
Spasms of stomach, with griping: *Magnes. Phos.*
Stomach sore to touch: *Calc. Phos.*
" tender to touch: *Ferrum Phos.*
Stomachache accompanied by:
" " constipation: *Kali Mur.*
" " depression: *Kali Phos.*
" " exhaustion: *Kali Phos.*
" " loose evacuations: *Ferrum Phos.*
" from acidity of the stomach: *Natr. Phos.*
" " chill: *Ferrum Phos.*
" " worms: *Natr. Phos.*
Temperature, rise of, in evening: *Kali Sulph.*
Thirst, great: *Natr. Mur.*
Thirstlessness: *Kali Phos.*
Vomiting, after cold drinks: *Calc. Phos.*
" bile: *Natr. Sulph.*
" bright-red blood: *Ferrum Phos.*
" dark, clotted blood: *Kali Mur., Ferrum Phos.*
" fluids like coffee-grounds: *Natr. Phos.*
" from stomachache: *Magnes. Phos.*
" greenish water: *Natr. Sulph.*
" sour acid fluids: *Natr. Phos.*
" thick white phlegm: *Kali Mur.*
" undigested food: *Ferrum Phos., Calc. Fluor.*
" watery: *Natr. Mur.*
Water-brash, with constipation: *Natr. Mur.*
Water gathers in mouth: *Natr. Mur.*

ABDOMEN

Abdomen, bloated: *Kali Sulph., Magnes. Phos.*
" cold to touch: *Kali Sulph.*
" cutting pains in: *Natr. Sulph., Magnes. Phos., Ferrum Phos.*
" distended: *Magnes. Phos.*
" inflammation, fever: *Ferrum Phos.*
" sunken: *Calc. Phos.*
" swollen: *Kali Phos., Kali Mur.*
" tender to touch: *Kali Mur.*
Anus, itching at: *Natr. Phos., Calc. Fluor.*
" cracks and fissures of the: *Calc. Phos.*
Anus, pain in: *Kali Mur.*
Back, pain in: *Calc. Fluor.*
Belching: *Magnes. Phos.*
Bilious evacuations: *Natr. Sulph.*
Bowels, loose, in old people: *Natr. Sulph.*
" sore and tender: *Ferrum Phos.*
Burning in the bowels: *Silicea.*
" sore pain in the pit of the stomach: *Ferrum Phos.*
Burning pain in rectum: *Natr. Mur.*
Colic of infants: *Magnes. Phos.*
Constant urging to stool: *Kali Mur.*
Constipation, see Stools.
Diarrhea, see Stools.
Distended abdomen: *Magnes. Phos.*
Feces, inability to expel: *Calc. Fluor.*
Flatulence, with pains in left side: *Kali Phos.*
Flatulent colic: *Natr. Phos., Natr. Sulph.*
Flatulent distention of the abdomen: *Natr. Mur.*

Frequent calls to stool, no passage: *Calc. Phos., Kali Phos., Magnes. Phos.*
Gnawing in bowels: *Magnes. Phos.*
Heartburn: *Ferrum Phos.*
Heat in lower bowels: *Natr. Sulph., Ferrum Phos.*
Infant cries when it nurses: *Calc. Phos.*
Liver, pains in region of: *Kali Mur.*
" sensitive: *Natr. Sulph.*
" sharp, shooting pains in: *Natr. Sulph.*
" sluggish: *Kali Mur.*
" region of, sore to touch: *Natr. Sulph.*
Neuralgia of bowels: *Magnes. Phos.*
" rectum: *Calc. Phos.*
Pain of a colicky nature caused by sudden change from hot to cold: *Kali Sulph.*
Pains in abdomen relieved by pressure: *Magnes. Phos.*
" " rubbing: *Magnes. Phos.*
" " warmth: *Magnes. Phos.*
Rectum, pain in: *Magnes. Phos.*
Right shoulder-blade, pains under: *Kali Mur.*
Sluggish action of the liver, with pale yellow evacuations; pain in region of liver or under the right shoulder-blade: *Kali Mur.*
" constipation, with furred tongue and protruding eyeballs: *Kali Mur.*
Spasmodic pains: *Magnes. Phos.*
Sulphurous odor of gas from bowels: *Kali Sulph.*
Swelling of abdomen: *Kali Mur.*
Torn feeling after stool: *Natr. Mur.*
Vomiting of bile: *Natr. Sulph.*
" curdled masses: *Natr. Phos.*

STOOLS

Bowels discharging mattery substances: *Calc. Sulph.*

Constipation from dryness of the mucous membrane with watery secretions in other parts: *Natr. Mur.*

" light colored stool, showing want of bile: sluggish action of the liver: *Kali Mur.*

" with drowsiness and watery symptoms from the eyes or mouth: *Natr. Mur.*

" " dull, heavy headache, profusion of tears or vomiting of frothy mucus: *Natr. Mur.*

Diarrhea after eating greasy, fatty food: *Kali Mur.*

" alternating with constipation: *Natr. Mur.*

" especially of children, with green, sour smelling stools caused by an acid condition: *Natr. Phos.*

" in teething children; stools slimy, green, undigested, with colic: *Calc. Phos.*

" like water: *Natr. Mur.*

" of school girls, accompanied by headache: *Calc. Phos.*

" stools frothy, slimy, causing soreness and smarting: *Natr. Phos.*

" when there is much straining at stool or constant urging to stool with passing of jelly-like mucus indicating acidity: *Natr. Phos.*

" with greenish, bilious stools or vomiting of bile: *Natr. Sulph.*

" " pale, yellow, clay-colored stool, swelling of the abdomen, slimy stools: *Kali Mur.*

Diarrhea, putrid, foul evacuations, depression and exhaustion of the nerves: *Kali Phos.*
" yellow, slimy, purulent matter: *Kali Sulph.*
Flatulent colic with green sour smelling stools, or vomiting of curdled masses: *Natr. Phos.*
Frequent call for stool, but passes nothing: *Calc. Phos.*
Griping pain in the abdomen with watery diarrhea, stools expelled with force: *Natr. Mur.*
Loose morning stool, worse in cold, wet weather: *Natr. Sulph.*
Looseness of the bowels in old people: *Natr. Sulph.*
" with watery stools: *Natr. Mur.*
Offensive stools: *Calc. Phos., Kali Phos.*
Retention of stool: *Natr. Mur.*
Stool is hot, often noisy and offensive: *Calc. Phos.*
Stools are dry and often produce fissures in the rectum: *Natr. Mur.*

URINARY ORGANS

Albuminous urine: *Kali Phos., Calc. Phos.*
Bladder, chronic inflammation of: *Kali Mur., Ferrum Phos., Calc. Sulph.*
Brickdust sediment in urine: *Natr. Sulph.*
Burning after urinating: *Natr. Mur., Ferrum Phos.*
" pain over kidneys: *Ferrum Phos.*
Constant urging to urinate, if not chronic: *Ferrum Phos.*
Cutting pains after urinating: *Natr. Mur.*
" pains at neck of bladder: *Calc. Phos.*
Dark red urine with rheumatism: *Natr. Phos.*
Desire to urinate with scanty emission: *Silicea.*

Enuresis of children (wetting of the bed): *Kali Phos., Natr. Phos., Ferrum Phos.*
" if from worms: *Natr. Phos.*
Excessive flow of watery urine: *Natr. Mur., Ferrum Phos.*
First stage of inflammation of the bladder causing retention of urine with pain and smarting when urinating: *Ferrum Phos.*
Frequent urging to urinate with sharp shooting pains, cutting at the neck of the bladder and along the urethra: *Calc. Phos.*
" urination with inability to retain the urine, and acidity: *Natr. Phos.*
" passing of much water, sometimes scalding: *Kali Phos.*
Gravel in bilious persons: *Natr. Sulph.*
" pain while passing: *Natr. Sulph., Magnes. Phos.*
" sediment in urine: *Natr. Sulph., Calc. Phos.*
" with gouty symptoms: *Natr. Sulph.*
Great thirst, with excessive flow of watery urine: *Natr. Mur.*
Highly-colored urine: *Calc. Phos., Ferrum Phos.*
" with fever: *Natr. Phos., Ferrum Phos.*
Inability to retain urine, from nervous debility: *Kali Phos.*
Incontinence, weakness of sphincter: *Ferrum Phos.*
Increase in quantity of urine: *Calc. Phos.*
Increased quantity of urine when traced to a relaxed condition of the muscle fibers of the urinary organs: *Calc. Fluor.*
Involuntary emission of urine while walking: *Natr. Mur.*

Pain when passing gravel: *Magnes. Phos.*

Sandy deposits in urine: *Natr. Sulph.*

Sediment clings to side of vessel: *Natr. Sulph.*

Sharp shooting pains at neck of bladder: *Calc. Phos.*

Smarting on urinating: *Ferrum Phos.*

Spasmodic retention of urine: *Magnes. Phos.*

Spasms of bladder, with painful straining: *Magnes. Phos., Ferrum Phos., Kali Phos.*

Urine frequently scalding: *Kali Phos.*

" suppression of: *Ferrum Phos.*

" dark colored, when there is torpidity and inactivity of the liver: *Kali Mur.*

" high colored: *Calc. Phos.*

" " " feverish: *Ferrum Phos.*

FEMALE ORGANS

Abdominal spasms followed by leucorrhea: *Magnes. Phos.*

Aching in uterus: *Calc. Phos.*

Acid leucorrhea, worse after menstruating: *Calc. Phos.*

Acrid pain during leucorrhea, with yellowness of the face: *Natr. Mur.*

After confinement when the pelvic muscles are relaxed: *Calc. Fluor.*

Colic in nervous, lachrymose women: *Kali Phos., Magnes. Phos.*

Congestion of the uterus, menstrual periods too frequent; lasting too long: *Kali Mur.*

Discharge, deep-red or blackish-red: *Kali Phos.*
" scalding, smarting: *Natr. Mur.*
" sickening: *Natr. Phos.*
" sour-smelling: *Natr. Phos.*
" thick, white, bland: *Kali Mur.*
Discharge, thin, with offensive odor: *Kali Phos.*
Dragging in groin: *Calc. Fluor.*
" " small of back: *Calc. Fluor.*
Dryness of vagina: *Natr. Mur.*
Dysmenorrhea: *Magnes. Phos.*
" labor-like pains during: *Magnes. Phos.*
Dysmenorrhea with congestion: *Ferrum Phos.*
" vomiting of undigested food: *Ferrum Phos.*
Increased menses: *Silicea.*
Leucorrhea, accompanied by:
" albuminous discharge: *Calc. Phos.*
" milky-white, non-irritating discharge: *Kali Mur.*
" rawness and itching of parts: *Natr. Phos.*
" scalding, acrid discharge: *Natr. Mur., Kali Phos.*
" slimy, greenish discharge: *Kali Sulph.*
" thick, yellow, bloody discharge: *Calc. Sulph.*
" watery, slimy, excoriating discharge: *Natr. Mur.*
" yellow, creamy discharge: *Natr. Phos.*
Menstruation, accompanied by:
" acrid leucorrhea: *Calc. Phos., Natr. Phos.*
" bearing-down pains: *Calc. Fluor.*
" cold extremities: *Calc. Phos., Ferrum Phos.*
" colic: *Natr. Sulph., Kali Phos., Magnes. Phos.*
" constipation: *Silicea, Natr. Sulph.*
" excitableness: *Kali Phos.*

Menstruation, accompanied by:
" flushed face: *Ferrum Phos., Calc. Phos.*
" fullness in abdomen: *Kali Sulph.*
" headache: *Natr. Mur., Ferrum Phos., Kali Phos.*
" hysteria: *Kali Phos.*
" icy coldness of body: *Silicea.*
" labor-like pains: *Calc. Phos., Magnes. Phos.*
" morning diarrhea: *Natr. Sulph.*
" nervousness: *Kali Phos.*
" pains in back: *Calc. Phos.*
" sadness: *Natr. Mur.*
" watery leucorrhea: *Natr. Mur.*
" weeping: *Natr. Mur.*
" weight in abdomen: *Kali Sulph.*
Menstruation, delayed, in young girls: *Natr. Mur.*
" retarded: *Kali Mur.*
" thin, watery blood: *Natr. Mur., Kali Phos.*
" too frequent: *Calc. Phos.*
" " late: *Natr. Mur., Kali Sulph., Kali Phos.*
" " profuse: *Kali Mur., Calc. Fluor., Kali Phos.*
Menstrual flow bright-red blood: *Ferrum Phos.*
" dark, clotted, black blood: *Kali Mur.*
" stringy and fibrous: *Magnes. Phos.*
Menstruations of pale, nervous, sensitive women: *Kali Phos.*
Neuralgia of the ovaries: *Magnes. Phos.*
Pains precede monthly flow: *Magnes. Phos.*
Sticking pains in vagina: *Natr. Mur.*
Thighs, pain extends to: *Calc. Fluor.*
Vagina, smarting of, after urinating: *Natr. Mur.*
Vaginal secretions, acid: *Natr. Phos.*
" " watery, creamy, yellow: *Natr. Phos.*

RESPIRATORY ORGANS

Aching in chest: *Calc. Phos.*

Acute inflammation of the windpipe with expectoration of frothy, watery mucus, constant frothy expectoration: *Natr. Mur.*

Acute, painful, short, irritating cough: *Ferrum Phos.*

All inflammatory conditions of the respiratory tract, in the first stage: *Ferrum Phos.*

All symptoms worse in damp weather, also rainy weather: *Natr. Sulph.*

Breath, short of, from asthma, worse from motion: *Kali Phos.*

Breathing, hurried, at beginning of disease: *Ferrum Phos.*

Catch in breath: *Ferrum Phos.*

Children, cough of teething: *Calc. Phos.*

Chronic coughs: *Calc. Phos.*

Cold in chest: *Ferrum Phos.*

Constant spitting of frothy water: *Natr. Mur.*

Constriction of chest: *Magnes. Phos.*

Convulsive fits of coughing: *Magnes. Phos.*

Cough, better in cool open air: *Kali Sulph.*

" hard, dry: *Ferrum Phos.*

" irritating, painful: *Ferrum Phos.*

" pain in chest from: *Natr. Mur., Ferrum Phos.*

" with headache: *Natr. Mur.*

" " hectic fever: *Calc. Sulph.*

" " mattery expectoration: *Calc. Sulph.*

" worse in evening: *Kali Sulph.*

" " warm room: *Kali Sulph.*

Countenance, pale livid: *Kali Phos.*

Croupy hoarseness: *Kali Mur., Kali Sulph.*
Expectoration, albuminous: *Calc. Phos.*
" difficult: *Natr. Mur., Kali Mur.*
" salty: *Natr. Mur.*
" slips back: *Kali Sulph.*
" streaked with blood: *Ferrum Phos.*
" thick, yellow, green: *Silicea.*
" tiny yellow lumps: *Calc. Fluor., Silicea.*
" watery: *Natr. Mur.*
" yellow, green, slimy: *Kali Sulph.*
Harsh breathing: *Natr. Sulph.*
Hawking, to clear throat: *Calc. Phos.*
Hoarseness from cold: *Kali Mur.*
" " overexertion of voice: *Calc. Phos.*
" of speakers: *Ferrum Phos.*
Inflammatory condition of the respiratory tract when the expectoration is decidedly yellowish, greenish and slimy: *Kali Sulph.*
Loss of voice: *Kali Mur.*
" from paralysis of the vocal cords: *Kali Phos.*
Loud, noisy cough: *Kali Mur.*
Must go in open air for relief: *Kali Sulph.*
" sit up: *Magnes. Phos.*
Nervous depression: *Kali Phos.*
Painful hoarseness and huskiness of speakers and singers when due to irritating bronchi: *Ferrum Phos.*
Râles in chest: *Kali Mur.*
Rattling in chest: *Kali Mur., Natr. Mur.*
Sharp pains in chest: *Magnes. Phos.*
Short spasmodic cough like whooping cough: *Kali Mur.*

Shortness of breath from asthma or with exhaustion or want of proper nerve power; worse from motion or exertion: *Kali Phos.*
Soreness of chest: *Ferrum Phos.*
Spasmodic cough: *Magnes. Phos., Kali Phos., Kali Mur.*
" at night: *Magnes. Phos.*
" worse lying down: *Magnes. Phos.*
Sudden, shrill voice: *Magnes. Phos.*
Suffocates in heated room: *Kali Sulph.*
Stomach cough: *Kali Mur.*
Thick, tenacious, white phlegm: *Kali Mur.*
Threatened suffocation: *Kali Phos.*
Tickling in throat: *Calc. Fluor.*
Tongue coated white: *Kali Mur.*
Weakness and prostration: *Calc. Phos.*

NERVOUS SYMPTOMS

Coldness: *Calc. Phos.*
" after attack of nervousness: *Kali Phos.*
Colic, worse at night: *Calc. Phos.*
Convulsions: *Calc. Phos.*
" of teething children: *Ferrum Phos., Calc. Phos., Magnes. Phos.*
Cramps in limbs: *Magnes. Phos.*
" worse at night: *Calc. Phos.*
Creeping numbness: *Calc. Phos.*
Cries easily: *Kali Phos.*
Despondent: *Kali Phos.*
Dwells upon grievances: *Kali Phos.*
Feels pain keenly: *Kali Phos.*
Feet twitch during sleep: *Natr. Sulph.*

Grinding of teeth, from worms: *Natr. Phos.*
Hands twitch during sleep: *Natr. Sulph.*
Head, involuntary shaking of: *Magnes. Phos.*
Hysteria: *Natr. Mur., Kali Phos.*
Impatient: *Kali Phos.*
Involuntary motion of hands: *Magnes. Phos.*
Irritable: *Kali Phos.*
Lassitude: *Natr. Sulph.*
Nervous sensitiveness: *Kali Phos.*
Neuralgia accompanied by:
" congestion, after taking cold: *Ferrum Phos.*
" depression: *Kali Phos.*
" failure of strength: *Kali Phos.*
" flow of saliva: *Natr. Mur.*
" " tears: *Natr. Mur.*
" shifting pains: *Kali Sulph.*
" " in any organ: *Kali Phos.*
Neuralgia, obstinate, heat or cold gives no relief: *Silicea.*
" " occurring at night: *Silicea, Calc. Phos.*
" periodic: *Magnes. Phos., Natr. Mur.*
" relieved by gentle motion: *Kali Phos.*
" " pleasant excitement: *Kali Phos.*
" sensitive to light and noise: *Kali Phos.*
" worse at night: *Kali Phos.*
" " in cold weather: *Natr. Mur.*
" " " the morning: *Natr. Mur.*
" " when alone: *Kali Phos.*
Pains, like electrical shocks: *Calc. Phos., Magnes. Phos.*
" " trickling of cold water: *Calc. Phos.*
Sensibility, want of: *Magnes. Phos.*

Spasms occurring at night: *Silicea, Magnes. Phos., Calc. Phos.*

Tired, weary and exhausted: *Calc. Phos., Magnes. Phos., Kali Phos.*

Tired, weary, with biliousness: *Natr. Sulph.*

Trembling hands: *Magnes. Phos.*

Writer's cramp: *Magnes. Phos.*

SKIN SYMPTOMS

Abscess, for heat and pain: *Ferrum Phos.*

Albuminous discharge: *Calc. Phos.*

Blisters, with fetid, watery contents: *Kali Phos.*

" with clear watery contents: *Natr. Mur.*

Burns, when suppurating: *Calc. Sulph.*

Burning, as from nettles: *Calc. Phos.*

Chafed skin of infants: *Natr. Phos., Natr. Mur.*

Chapped hands from cold: *Ferrum Phos., Calc. Fluor.*

Chilblains: *Kali Phos., Kali Mur.*

Colorless, watery vesicles: *Natr. Mur.*

Cracks in palms of hands: *Calc. Fluor.*

Dandruff: *Kali Sulph., Natr. Mur., Kali Mur.*

Discharge, albuminous: *Calc. Phos.*

" blood and pus: *Calc. Sulph.*

Discharge, fetid: *Kali Phos.*

" thick, yellow pus: *Silicea.*

Dry skin: *Calc. Phos., Kali Sulph.*

Eruptions with watery contents: *Natr. Mur.*

" thick, white contents: *Kali Mur.*

Excessive dryness of skin: *Natr. Mur.*

Exudations, when white and fibrinous indicate a deficiency in: *Kali Mur.*
" albuminous: *Calc. Phos.*
" yellow with small, tough lumps: *Calc. Fluor.*
" " like gold: *Natr. Phos.*
" yellowish and slimy or watery: *Kali Sulph.*
" greenish, thin: *Kali Sulph.*
" clear, transparent, thin like water: *Natr. Mur.*
" mattery, or streaked with blood: *Calc. Sulph.*
" pus is thick, yellow: *Silicea.*
" very offensive smelling: *Kali Phos.*
" causing soreness and chafing: *Natr. Mur., Kali Phos.*
Face full of pimples: *Calc. Phos.*
Festers easily: *Calc. Sulph., Silicea.*
Fevers, with skin dry and hot: *Kali Sulph.*
Freckles: *Calc. Phos.*
Greasy scales on skin: *Kali Phos.*
Hard, callous skin: *Calc. Fluor.*
Heals slowly: *Silicea.*
Herpetic eruptions: *Natr. Mur.*
Hives: *Natr. Phos.*
Horny skin: *Calc. Fluor.*
Inflammation of skin, for fever and heat: *Ferrum Phos.*
" with yellow, watery exudation: *Natr. Sulph.*
Irritating secretions: *Kali Phos.*
Irritation of the skin similar to chilblains: *Kali Mur.*
Itching, as from nettles: *Calc. Phos.*
" of skin, with crawling: *Kali Phos., Calc. Phos.*
" without eruptions: *Calc. Phos.*
Mattery scabs on heads of pimples: *Calc. Sulph.*
Moist scabs on skin: *Natr. Sulph.*

Nettle-rash, after becoming overheated: *Natr. Mur.*

Perspiration, lack of: *Kali Sulph.*

" to promote: *Kali Sulph.*

Pimples all over body, like flea-bites: *Natr. Phos.*

" with itching: *Calc. Phos.*

" under beard: *Calc. Sulph.*

Pustules on face: *Silicea, Kali Mur.*

Rawness of skin in little children: *Natr. Phos.*

Scalds, when suppurating: *Calc. Sulph.*

Scaling eruptions on skin: *Calc. Phos.*

Scrofulous eruptions: *Silicea, Calc. Phos.*

Secretions irritate: *Kali Phos.*

Shingles: *Natr. Mur., Kali Mur.*

Skin affections with vesicular eruptions containing yellowish water: *Natr. Sulph.*

" yellow scabs: *Calc. Sulph.*

Skin festers easily: *Calc. Sulph.*

" hard and horny: *Calc. Fluor.*

" heals slowly and suppurates easily after injuries: *Silicea.*

" dry, hot and burning. lack of perspiration: *Kali Sulph.*

" itching and burning; as from nettles: *Calc. Phos.*

" scales freely on a sticky base: *Kali Sulph.*

" withered and wrinkled: *Kali Phos.*

Suppurates easily: *Silicea.*

To aid desquamation in eruptive diseases: *Kali Sulph.*

To assist in the formation of new skin: *Kali Sulph.*

Ulcers around nails: *Silicea.*
" fistulous, thick, yellow pus: *Silicea, Calc. Fluor., Calc. Sulph.*
Unhealthy-looking skin: *Silicea.*
Yellow scabs: *Calc. Sulph.*
" scales on skin: *Natr. Sulph.*
Warts: *Kali Mur.*
" in palms of hands: *Natr. Mur.*
Wounds do not heal readily: *Calc. Sulph.*
" neglected, discharge pus: *Calc. Sulph.*
Wrinkled skin: *Kali Phos.*

FEBRILE (FEVER) SYMPTOMS

Acid symptoms during fever: *Natr. Phos.*
Bilious fevers: *Natr. Sulph.*
Catarrhal fever, chilly sensations: *Ferrum Phos.*
" quickened pulse: *Ferrum Phos.*
Chill commencing in the morning about 10 o'clock and continuing till noon, preceded by intense heat, increased headache and thirst, sweat, weakening, more backache and headache; great languor, emaciation, sallow complexion and blisters on the lips; *Natr. Mur.*
Chilliness at beginning of fevers: *Ferrum Phos.*
" in back: *Natr. Mur.*
Chills run up and down spine: *Magnes. Phos.*
Clammy sweat on body: *Calc. Phos.*
Cold sweat on face: *Calc. Phos.*
Copious night sweat with prostration: *Silicea.*
Drowsiness: *Natr. Mur.*
Dull, heavy headache: *Natr. Mur.*

Excessive exhausting perspiration or sweating, while eating, with weakness at the stomach: *Kali Phos.*

Feeling of chilliness especially in the back; watery saliva; full heavy headache, increased heat: *Natr. Mur.*

Fevers, during suppurative processes: *Silicea.*

" vomit of sour fluids during: *Natr. Phos.*

" with chills and cramps: *Magnes. Phos., Ferrum Phos.*

First stage of fevers: *Ferrum Phos.*

Flashes of heat from indigestion: *Natr. Phos.*

Frontal headache from flashes of heat: *Natr. Phos.*

Gastric fever, first stage: *Ferrum Phos.*

" when the temperature runs in the evening: *Kali Sulph.*

Increased thirst: *Natr. Mur.*

Inflammations, first stage: *Ferrum Phos.*

" second stage: *Kali Mur.*

In eruptive fevers to aid desquammation: *Kali Sulph.*

Living in damp regions: *Natr. Mur.*

Much sweat in the daytime: *Natr. Mur.*

Nervous chills, with chattering of teeth: *Magnes. Phos., Kali Phos.*

Nervous fever: *Kali Phos.*

Night sweats: *Silicea.*

Offensive foot-sweats: *Silicea.*

Perspiration, excessive: *Calc. Phos., Kali Phos.*

" sour-smelling: *Natr. Phos.*

Profuse night-sweats: *Natr. Mur., Silicea, Calc. Phos.*

Pulse, subnormal: *Kali Phos.*

Rheumatic fever, heat and congestion: *Ferrum Phos.*
Saliva clear, watery: *Natr. Mur.*
Second stage of fever: *Kali Mur.*
Shivering at beginning of fever: *Calc. Phos., Ferrum Phos.*
Sleeplessness: *Kali Phos.*
Stupor: *Natr. Mur., Kali Phos.*
Sweat of head in children: *Silicea.*
" while eating: *Kali Phos.*
To assist in promoting perspiration: *Kali Sulph.*
Tongue coated dirty, greenish-brown: *Natr. Sulph.*
" grayish-white, slimy: *Kali Mur.*
Twitching: *Natr. Mur.*

SLEEP

Better in evening: *Natr. Sulph.*
Congestion of blood to the head at night: *Silicea.*
Constant desire to sleep in morning: *Natr. Mur.*
Drawing pain in the back at night during sleep: *Natr. Mur.*
Dreams much: *Natr. Sulph.*
Drowsiness, with bilious symptoms: *Natr. Sulph.*
Drowsy: *Natr. Mur., Magnes. Phos., Calc. Phos.*
Dull: *Natr. Mur.*
Empty feeling in stomach: *Kali Phos.*
Frequent dreams and exclamations during sleep: *Silicea.*
Great drowsiness: *Silicea.*
Grits teeth: *Natr. Phos.*
Hard to wake in morning: *Calc. Phos.*
Heavy, anxious dreams: *Natr. Sulph.*

Jerking of limbs during sleep: *Silicea, Natr. Sulph.*

Much yawning: *Silicea.*

Nightmare, with bilious symptoms: *Natr. Sulph., Kali Sulph.*

Picks nose: *Natr. Phos.*

Restless sleep, from worms: *Calc. Phos., Natr. Phos.*

Screams in sleep: *Natr. Phos.*

Sleep does not refresh: *Natr. Mur.*

Sleepiness, with hectic fever: *Calc. Sulph.*

Sleeplessness, after excitement: *Ferrum Phos., Natr. Phos.*

" from nervous causes: *Kali Phos.*

" " worry: *Ferrum Phos., Kali Phos.*

Sleepy in morning: *Natr. Sulph.*

Stretching, from nervous causes: *Kali Phos.*

Stupid: *Natr. Mur.*

Tired in morning: *Natr. Mur., Natr. Sulph.*

Vivid dreams: *Calc. Phos.*

Wakefulness: *Kali Phos., Ferrum Phos.*

Wakefulness in old people: *Silicea.*

Weariness, from nervous causes: *Kali Phos.*

" with bilious symptoms: *Natr. Sulph.*

Yawning, from nervous causes: *Kali Phos.*

" with spasmodic straining of lower jaw: *Natr. Phos.*

THE INORGANIC SALTS

Copied from "A Text-Book of Physiology."

By William H. Howell, Ph.D., M.D., Sc.D., LL.D.,
Professor of Physiology in The Johns Hopkins University, Baltimore.

"The body contains in its tissues and liquids a considerable amount of inorganic material. When any organ is incinerated this material remains as the ash. If we include the bones, which are rich in mineral matter, the average amount of ash in the body amounts to about 4.3 to 4.4 per cent of its weight. The bones, however, in the adult contain most of this ash (five-sixths). In the soft tissues, like the muscle, the ash constitutes about 0.6 to 0.8 per cent of the moist weight. The ash consists of chlorids, phosphates, sulphates, carbonates, fluorids, or silicates of potassium, sodium, calcium, magnesium, and iron; iodin occurs also, especially in the thyroid tissues.

"In the liquids of the body the main salts are sodium chlorid, sodium carbonate, sodium phosphate, potassium and calcium chlorid or phosphate. In considering the organic foodstuffs weight was laid upon their value as sources of energy, as well as upon their function in constructing tissue. The salts have no importance from the former standpoint. Whatever chemical changes they undergo are not attended by any liberation of heat energy—none at least of sufficient importance to be considered.

"They have, however, most important functions. They maintain a normal composition and osmotic pressure in the liquids and tissues of the body, and by virtue of their osmotic pressure they play an important part in controlling the flow of water to and from the tissues. Moreover, the salts constitute an essential part of the composition of living matter. In some way they are bound up in the structure of the living molecule and are necessary to its normal reactions or irritability. Even the proteins of the body liquids contain definite amounts of ash, and if this ash is removed their properties are seriously altered, as is shown by the fact that ash-free native proteins lose their property of coagulation by heat. The globulins are precipitated from their solutions when the salts are removed.

"The special importance of the calcium salts in the coagulation of blood and the curdling of milk has been referred to, as also the peculiar part played by the calcium, potassium, and sodium salts in the rhythmical contractions of heart muscle, the irritability of muscular and nervous tissues, and the permeability of the capillary walls and other membranes. The special importance of the iron salts for the production of hemoglobin is also evident without comment.

"There can be no doubt, in fact, that each one of the salts of the body has a special nutritive value and a special metabolic history. The time will doubtless come when the special importance of the potassium, sodium, calcium, and magnesium will be understood as well, at least, as we now understand the significance of iron, and quite possibly this knowledge will find a direct therapeutic application as in the case of iron."

INDEX

A

B

C

D

E

F

G

H

I

J

K

L

M

N

P

R

R

S

T

U

V

W